D1391250

BACTERIOLOGY
ILLUSTRATED

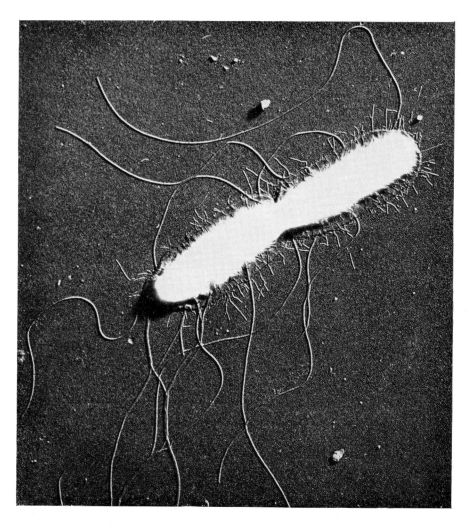

Electron micrograph of *Salmonella typhi* ×16,000. The dividing bacillus possesses 12 flagella and approximately 100 fimbriae.

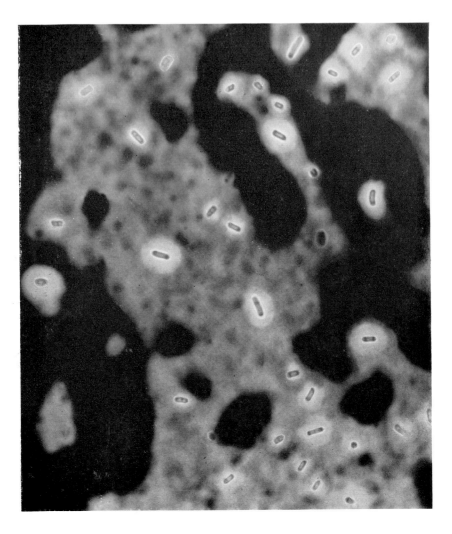

Klebsiella aerogenes strain A3, capsular serotype 54 grown for 7 days on maltose peptone agar. Capsules and loose slime in wet film with India ink. ×2000

Within each large capsule lies a central protoplast surrounded by a narrow, bright diffraction halo; the capsulate cells lie on a background of loose slime which is slightly darker than the capsule since it is partly and patchily infiltrated by the carbon particles of the ink. In contrast to the capsules the variety of shapes and arrangements of loose slime should be noted.

BACTERIOLOGY
ILLUSTRATED

R. R. GILLIES
M.D., F.R.C.P.E., D.P.H., M.R.C.Path.

Reader in Bacteriology, University of Edinburgh

T. C. DODDS
F.I.M.L.T., F.I.I.P., F.R.P.S.

Late Director of the Medical Photography Unit, University of Edinburgh

Foreword by
R. CRUICKSHANK
C.B.E., M.D., F.R.C.P., D.P.H., F.R.S.E.

Formerly Professor of Bacteriology, University of Edinburgh

THIRD EDITION

CHURCHILL LIVINGSTONE
EDINBURGH AND LONDON
1973

First Edition 1965
Second Edition 1968
Third Edition 1973
Italian Edition 1970
German Edition 1973

ISBN 0 443 01013

Printed in Great Britain

FOREWORD

IN the learning process a combination of seeing and hearing is usually more effective than either seeing or hearing alone. The blackboard, the diagram, the lantern slide and the ciné film are all useful aids to the teacher in imparting the great mass of factual data which the student requires to memorise if he is to satisfy the examiners. In the field of medicine the student's burden of necessary knowledge becomes ever heavier and methods for easing the load have not kept pace with the accumulation of new information. In the Departments of Pathology and Bacteriology of the Edinburgh University Medical School informative pictures and drawings have for long played an important part in the processes of teaching and learning, largely under the influence of Richard Muir, a superb technician and expert illustrator who later handed on the torch to his disciple, Tom Dodds.

It is thirty-seven years since Richard Muir first produced his Bacteriological Atlas and I confess unashamedly to having mutilated a copy in order to prepare a set of passe-partout framed pictures to hang on the walls of the practical bacteriology classroom. With the developments of colour photography and a great deal of enthusiastic effort Cranston Low and Dodds produced in 1947 a successor to Muir's Atlas, which illustrated with remarkable fidelity many of the microscopic appearances and, perhaps less successfully, some of the cultural characteristics of the common micro-organisms.

Now the phoenix has risen again and Mr Dodds and Dr R. R. Gillies have produced a new bacteriological atlas which has a number of distinctive and attractive features. It is in the best sense an illustrated primer in bacteriology and should prove most useful not only to the medical student in his efforts to grapple with the essentials of a rapidly expanding subject but also to students in other disciplines, including nursing, pharmacy and chiropody, who are expected to know some elementary microbiology. The last section should be particularly useful to the senior medical student, house officer and general practitioner, who need to know what kind of specimens are required for the laboratory diagnosis of infective syndromes such as sore throat, meningitis and urinary infections, and to have, also, some understanding of the steps and time involved in preparing an informative report. I know something of the care and effort that have gone into the building of this book and, symbolically breaking a bottle of stain over its covers, I wish it a most successful voyage.

Edinburgh, 1965   ROBERT CRUICKSHANK

PREFACE TO THIRD EDITION

THIS edition contains new sections on protozoa and fungi in an endeavour to make the volume more complete; the text has been revised and an enlarged bibliography is offered.

Fortune and misfortune often go hand in hand and the fortune here was that my friend T. C. Dodds had prepared much of the additional photographic material and proofed the blocks before his sudden death; therein lies the misfortune and I as well as many other medical authors have been deprived of the skills, advice, gentle criticism and friendship of a unique personality.

I have, however, had the unfailing support of Mr J. Paul, Acting Head of the Medical Photography Unit in the University, in completing this edition.

Several of the plates in the section on protozoa were prepared from material provided by Professor W. H. R. Lumsden; Dr A. C. Scott provided the material for the plates of *Cryptococcus neoformans*. Numerous colleagues have offered helpful advice and foremost among these are Professor J. P. Duguid and Dr J. G. Collee. Miss E. M. Clarke typed the script and gave much assistance in correcting the proofs.

Our engravers and Pillans and Wilson Ltd. have given unstinting support in producing the blocks and printing this edition and last but by no means least I would offer my sincere thanks to my publishers, and in particular to Mr W. G. Henderson and Mr A. D. Lewis.

R. R. GILLIES

Edinburgh, 1973

PREFACE TO FIRST EDITION

As bacteriology is essentially visual in its impact on the student, whether undergraduate, postgraduate or technical, and as most textbooks are inadequate in this respect, we have provided many photographic illustrations of actual microscopic and cultural appearances and only occasionally have artistic representations been employed. On the other hand, the text of this volume, by its brevity, pays tribute to the thorough verbal descriptions given in the many other books on the subject.

The first section of this work is synoptic in its approach but the bibliography should assist the reader in his search for authoritative information on the morphology, physiology and classification of bacteria. Only in this section have the illustrations been given figure numbers to facilitate cross-reference.

Many genera are dealt with systematically in Section II, which however is not exhaustive, but attempts to highlight the essential features of each genus which are necessary for the isolation and identification of species. Unless stated to the contrary, all illustrations of colonies are actual size.

In the last section the various procedures used in diagnostic bacteriology have been silhouetted and it is hoped that the processing charts, used in conjunction with relevant chapters in Section II, will help the reader to appreciate laboratory methodology. Thus we offer this volume as a *vade mecum* to lecture courses and practical classes; it should also be useful to the student preparing for examinations by providing him with a ready revision of much material which he has seen only once during his course of instruction.

We take responsibility for the content of this book but wish to acknowledge the ready assistance given by many colleagues. Professor J. P. Duguid kindly contributed the illustrations used for the Frontispiece and that on p. 4 for which we have also to acknowledge our gratitude to the Editors of the *Journal of General Microbiology* and the *Journal of Pathology and Bacteriology* respectively. Dr J. G. Collee prepared the material used to illustrate the phage-typing of a staphylococcus (page 42) and Nagler's reaction (page 81). Four consecutive undergraduate groups and also graduates studying for the Diploma in Public Health have, by their constructive criticisms, assisted us in adapting the text to a variety of needs. Many colleagues, both medical and technical, have also given us advice in this respect.

We must express our sincere thanks to Professor Robert Cruickshank who has written a most gracious foreword, to Mr D. Brown, Senior Technician, Department of Bacteriology, University of Edinburgh, to Mr J. Paul and Mr C. C. Allan of the Medical Photography Unit, University of Edinburgh, for invaluable assistance in the production of the many illustrations, to Scottish Studios and Engravers Co. Ltd. who produced the printing blocks, and by no means least, to our publishers, especially Mr Charles Macmillan, Mr J. Parker and Mr A. D. Lewis, for their ever ready help and guidance.

It is advised that, wherever possible, the colour illustrations be studied in tungsten light as the photographs were taken and the colour printing balanced under illumination of that type. Colour balance will appear inferior in daylight and fluorescent lighting.

THE AUTHORS

TABLE OF CONTENTS

SECTION IV—*PROTOZOOLOGY*

Chapter

SECTION V—*MYCOLOGY*

SECTION I
GENERAL INTRODUCTION

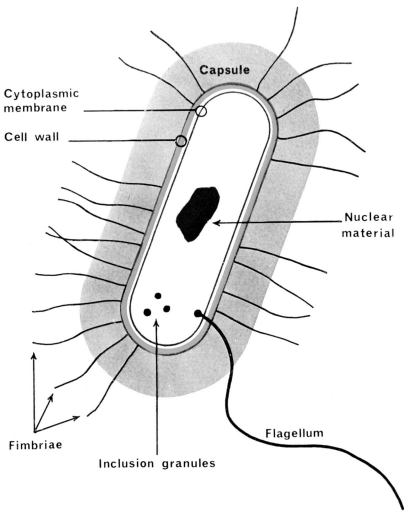

Capsule

Cytoplasmic
membrane

Cell wall

Nuclear
material

Fimbriae

Inclusion granules

Flagellum

Diagram of a typical bacterial cell.

CHAPTER 1

THE BACTERIAL CELL—MORPHOLOGY AND STAINING TECHNIQUES

BACTERIA have been described as 'bags of enzymes' but, in this chapter, we must take a glimpse at the anatomical features of the bacterial cell since these are essential to laboratory identification.

MORPHOLOGY

The size of bacteria. The unit of measurement is the micron $(\mu)=1/1000$ of a millimetre or 1/25,400 of an inch. Most cocci (spherical bacteria) are approximately 1μ in diameter and there is no virtue in memorising more detailed measurements for each type of coccus; in any case, such measurements have usually been made on killed and stained bacteria and thus may have little relationship to the living cell.

Bacilli (relatively straight, rod-shaped bacteria) show great variation in size, e.g., the anthrax bacillus (4 to 8μ by 1 to $1\cdot5\mu$) and the whooping-cough bacillus ($1\cdot5$ to $1\cdot8\mu$ by $0\cdot3$ to $0\cdot5\mu$). Such measurements are dependent on many environmental factors, e.g., the age of the cell, temperature during growth, availability of foodstuff, presence of antagonistic substances, etc.

With these facts in mind we can note certain physical features of a 'typical' bacterial cell.

The cell wall and cytoplasmic membrane. All bacteria are contained by a wall of semi-rigid materials that endows the cell with its particular shape; the cell wall is not essential to the survival of the bacterium and preparations free of cell walls (protoplasts) can be maintained in a carefully controlled environment.

Immediately within the cell wall lies the *cytoplasmic membrane*, composed essentially of lipo-protein, which selectively allows absorption of nutrients and excretion of waste products. This membrane, together with the cell wall, is probably involved in determining the response of the cell in Gram's staining reaction.

Nuclear apparatus. The presence or absence of a nucleus in bacteria has been argued for many years and continuation of such argument reflects the various definitions of the word 'nucleus'.

There can be no doubt that there is transfer of genetic information to daughter cells from the parent cell and for our purposes it matters little whether such information is contained in definable chromosomes; the nuclear apparatus differs from that of the eucaryotic cells of plants and animals in that it does not have a limiting membrane or nucleoli.

3

Intracellular granules. Various aggregations of material can be seen in many bacterial species and their composition can be judged by their affinity for certain dyes, e.g., lipid granules are recognised by their affinity for fat-soluble dyes.

The other commonly occurring type of granule is the volutin granule and this has a diagnostic significance in medical bacteriology; the demonstration of volutin granules assists differentiation of certain members of the genus *Corynebacterium* but it must be remembered that these granules will disappear from the bacterial cell when it is grown on mildly antagonistic media and reappear when subcultured on luxuriant media.

Two other types of intracellular granule have been noted but have, as yet, no direct application in medical bacteriology; they consist respectively of a starch-like polysaccharide and sulphur.

Capsules and extracellular slime. Bacteria which can form a thick gelatinous circumscribed layer outside the cell wall are described as capsulate; capsules develop readily when such bacteria are growing in host tissues but are maintained less readily or not at all on *in vitro* cultivation. Capsular substances are usually highly specific, chemically and immunologically, and allow type differentiation of otherwise identical species, e.g., pneumococci.

It should be appreciated that whilst certain disease-producing bacteria are pathogenic only in the capsulate state, capsules exist in saprophytic species so that capsulation is not necessarily equated with virulence.

Capsules usually are NOT revealed by ordinary staining methods and those of bacteria in cultures are most satisfactorily demonstrated with India ink. A loopful of India ink is mixed with a loopful of fluid culture or fleck of a surface colony; a coverslip is placed over the emulsion and pressed firmly down through a sheet of blotting paper. In contrast to the clearly delineated capsules, many capsulate and some non-capsulate species produce extracellular or loose slime; this colloidal material gives surface colonies a mucoid or sticky appearance and consistency. In India ink films, loose slime appears as irregular masses extending from the bacteria (Frontispiece).

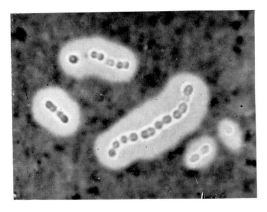

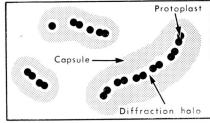

Wet India ink film of capsulate pneumococcus ×2000. The cocci are seen as faint grey areas within the clear capsules.

4

Flagella. With the exception of spirochaetes, motile strains of bacteria possess one or more filamentous appendages known as flagella; a flagellum is a long, thin filament twisted spirally, usually 0.02μ thick and longer than the bacterial cell from which it extrudes.

The distribution of flagella is constant in any one species and peritrichous and monotrichous arrangements are the most frequent in pathogenic species; flagella cannot be observed by ordinary microscopic techniques unless they are thickened by the deposition of stains on their surface. Electron microscopy gives a clear demonstration of flagella (Frontispiece).

FLAGELLA

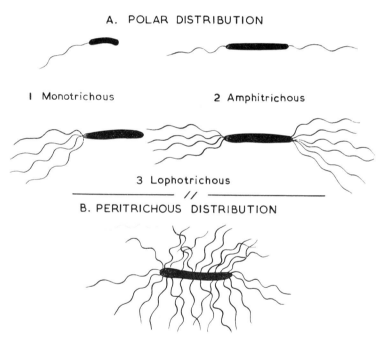

A. POLAR DISTRIBUTION

1 Monotrichous   2 Amphitrichous

3 Lophotrichous

———————— // ————————

B. PERITRICHOUS DISTRIBUTION

Diagram of various distributions of flagella.

For diagnostic purposes the presence of flagella is inferred by observing motility either in wet preparations examined microscopically or by noting the spreading growth which occurs when the organism is inoculated into a tube of semi-solid agar. The latter macroscopic method (page 6) is preferable since Brownian movement may be mistaken for motility in wet-film preparations, and also the intermittency of motility implies that without repeated and frequent observations a culture might be wrongly regarded as non-motile.

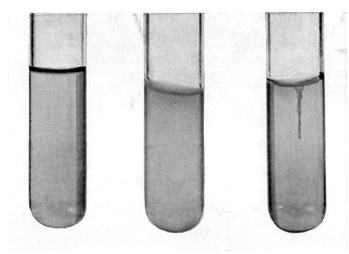

Macroscopic motility test. A pure culture is inoculated by means of a sterile straight wire into a tube of semi-solid agar to a depth of $\frac{1}{2}$–1 in. The tube on the left is uninoculated; the right-hand tube has been inoculated with a non-motile organism and after incubation growth is restricted to the inoculum track. The organisms inoculated into the middle tube were motile and they have spread throughout the medium; the inoculum track is not visible.

Furthermore, the preparation of satisfactory wet films is time-consuming and potentially dangerous to the technician and the alternative macroscopic method should thus encourage regular examination for motility without which the identification of many species, e.g., in the enterobacteria, will be delayed. The value of motility to the bacterial cell is not known although it is possible that motile species may have some advantage in spreading more rapidly through body fluids and tissues; motility may, by continuously changing the fluids in contact with the cell, assist the latter in the uptake of fresh foodstuffs.

Flagella have great practical importance to the bacteriologist since specific antisera can be prepared in animals and used for serological differentiation of various flagellar types; this is the basis of type-identification of the many members of the genus *Salmonella*.

Fimbriae. These are usually only half the width of flagella and are unrelated to motility; fimbriae can be seen only by electron microscopy (Frontispiece) but their presence can be implied by the haemagglutinating activity of certain bacterial species. Little is known of their function but since they occur in saprophytic and commensal bacteria as well as in pathogenic species they are probably unrelated to disease production. At present their importance lies in difficulties which they cause in diagnostic serology, e.g., antibodies to fimbriae have been noted in the sera of healthy persons and such sera have been found to react non-specifically in tests with diagnostic suspensions of salmonellae if the latter are fimbriate; thus such tests may be wrongly reported as positive.

Spores. These are highly resistant to adverse environmental conditions whether the latter are natural or artificially created by man; thus the spore can survive years of starvation in soil and, in comparison with the vegetative bacterial cell which is killed by exposure to 60°C for 10 min., spores are only destroyed by autoclaving at 100° to 120°C or higher for 10 or more min. Similarly spores are more resistant to disinfectants, drying and sunlight than are vegetative cells; spores have no reproductive significance, each vegetative cell forms only one spore and on germination the spore gives rise to a single vegetative cell.

The enhanced resistance of spores is due to the hard spore case, their low content of unbound water and very low metabolic activity as well as their high content of calcium and dipicolinic acid.

When spore-bearing organisms, e.g., those of the genus *Bacillus* or *Clostridium* are stained by Gram's method, the spores remain clear and un-stained; however, spores can be demonstrated by a modification of the Ziehl-Neelsen technique although their acid-fastness is less than that of tubercle bacilli.

The size, shape and position of the spore relative to the 'parent' vegetative cell are constant for any one bacterial species, thus by noting the particular combination of these features the species can usually be tentatively identified (page 8).

Spore preparations can be employed as a method of checking the efficiency of autoclaves and/or the operator; by including sealed envelopes of bacterial spores in different parts of the autoclave and the material being sterilised and, after autoclaving, attempting to recover viable cells, one can determine whether sterilisation has been effected. Strains commonly used for this purpose are *Bacillus subtilis* var. *globigii* (spores destroyed at 105°C in 15 min.) and *Bacillus stearothermophilus* whose spores require exposure to a temperature of 121°C for 10 to 30 min. for destruction.

Bacterial spores play an important part in the epidemiology of infections caused by members of the genus *Clostridium*, e.g., the spores of tetanus bacilli and *Cl. welchii* can remain viable in earth for many years after they have been produced from their respective vegetative cells and if introduced into a wound can then germinate with the production of fully virulent cells and cause serious if not fatal infections. Similarly in the case of anthrax, spores seeded on to pasture land or byres can survive and ultimately infect either man or other susceptible animals after long periods of time; this poses a special problem in prevention which is to some extent catered for by legislation regarding the disposal of carcases of animals dying from anthrax.

However, more recently, anthrax has become a domestic problem since fertiliser derived from animal bones may contain anthrax spores and this source has been incriminated in many cases of human anthrax in recent years.

SPORES

POSITION, SIZE AND SHAPE

GRAM'S STAIN   MODIFIED
ZIEHL - NEELSEN

Projecting, spherical and terminal

Non-projecting, ovoid and central

Non-projecting, ovoid and subterminal

Free spores

STAINING TECHNIQUES

Staining is of primary importance for the recognition of bacteria since their clear protoplasm is so feebly refractile that it is difficult to see them in the unstained condition unless dark-ground illumination is employed or, alternatively, use is made of phase-contrast techniques.

It is a prerequisite that the suspension to be stained must be firmly fixed to a glass slide before the various dye solutions are applied.

The making and fixing of a film—either a saline suspension of a colony or a sample from a fluid culture—is shown on page 10.

Gram's stain. This is the most commonly employed and important of all diagnostic staining techniques. By this method bacteria can be recognised as Gram-positive (blue-black) if they retain the primary dye complex of methyl violet and iodine in the face of attempted decolorisation or as Gram-negative (red) if decolorisation occurs as shown by the cell accepting the counterstain. The procedure in Gram's technique is shown on pages 12 and 13. It should be noted that any naturally Gram-positive organism may show a varying proportion of Gram-negative cells—these are dead cells which have lost the power of retaining the primary dye complex. The mechanism of Gram's staining method is not fully understood but may be associated with the cytoplasmic membrane and the presence of a magnesium ribonucleate-protein complex within cells which stain positively.

Ziehl-Neelsen's stain. There are three groups of acid-fast bacterial structures, certain bacteria, e.g., tubercle bacilli and leprosy bacilli, bacterial spores and lastly the 'clubs' associated with the presence in human or animal tissues of *Actinomyces*.

These groups are characterised by their relative resistance to ordinary dye-stuffs and by the fact that once they have been stained by the Ziehl-Neelsen technique (pages 14 and 15) they resist decolorisation by strong mineral acids.

Their degree of acid-fastness varies; tubercle bacilli will retain the red carbol-fuchsin stain when challenged with a 20 per cent. solution of H_2SO_4 whereas leprosy bacilli will only withstand decolorisation by 5 per cent. H_2SO_4. Spores tolerate only 0·25 to 0·5 per cent. H_2SO_4 and in the attempted decolorisation of actinomyces 'clubs' the strength of acid employed is 1 per cent.

MAKING AND FIXING FILM

1 STERILISE LOOP:
ALLOW TO COOL.

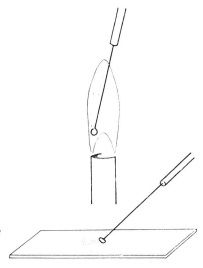

2 SPREAD LOOPFUL OF
MATERIAL ON SLIDE:
KEEP CLEAR OF EDGES.
RESTERILISE LOOP.

3 DRY FILM IN AIR OR
BY HOLDING <u>HIGH</u> OVER
BUNSEN FLAME.

4 FIX FILM ON SLIDE
BY SLOWLY PASSING IT
THREE TIMES THROUGH
BUNSEN FLAME: ALLOW
SLIDE TO COOL BEFORE
STAINING.

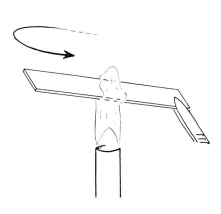

Albert's stain. Films are made and fixed and the steps in staining are given on page 16. In diagnostic bacteriology the demonstration of volutin granules is restricted to identification of diphtheria bacilli and the other members of the genus *Corynebacterium*, all of which are Gram-positive bacilli. It should be noted that many Gram-negative bacilli contain volutin granules so that an Albert-stained film must be evaluated in parallel with a Gram-stained film of the same preparation; even so the information thus obtained is only a part of the identification of a member of the genus *Corynebacterium*.

Many other staining techniques are used in bacteriological laboratories and details of these may be obtained in *Medical Microbiology* edited by Cruickshank (1968).

GRAM'S STAINING METHOD *

1 Flood slide with methyl-violet solution:
  Allow to act for 5 min.

Methyl-violet stain
1 per cent. aqueous solution of methyl-violet, 6B
                                        30 parts
5 per cent. solution of sodium bicarbonate
                                        8 parts

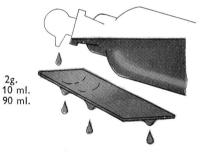

2 Wash off stain with iodine solution.

Iodine solution
Iodine   2g.
Normal solution of sodium hydroxide   10 ml.
Distilled water   90 ml.

3 Allow iodine to act for 2 min.

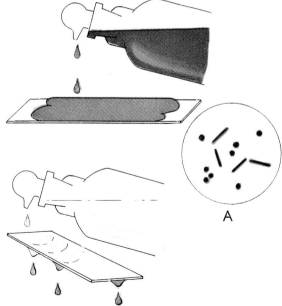

4 Drain off excess iodine. Decolorise with
  acetone for not more than 5 sec.

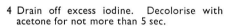

A

* As modified by Kopeloff, N., and Beerman, P
  (1922-23): *Proc. Soc. exp. Biol. (N.Y.)*, **20**, 71.

5 Wash slide immediately in water.

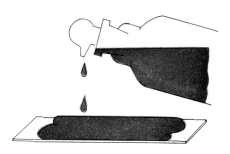

6 Apply basic fuchsin counterstain for 30 sec.

Basic fuchsin stain is a 0.05 per cent. aqueous solution of basic fuchsin.

7 Wash in water, blot and dry in air.

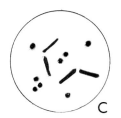

The inserts indicate the appearance of a mixed Gram +ve and Gram −ve film at different stages during staining.

A Before acetone decolorisation all organisms appear Gram +ve.

B After acetone decolorisation those organisms which are Gram −ve are no longer visible.

C These Gram −ve organisms are visualised after the application of the counterstain.

ZIEHL-NEELSEN'S STAINING METHOD

1 Flood slide with carbol fuchsin. Allow to act for 5 min. HEAT intermittently without boiling the stain.

Ziehl-Neelsen carbol fuchsin

Basic fuchsin 1 g.
Absolute alcohol 10 ml.
Phenol solution (5 per cent. in water) 100 ml.

The dye is dissolved in the alcohol and added to the phenol solution.

2 Wash with water.

3 Flood slide with 20 per cent. H_2SO_4. After 1 min. wash in water and apply fresh acid. Repeat process several times.

4 Wash thoroughly with water.

5 Apply 95 per cent. alcohol for 2 min.

6 Wash with water.

7 Apply methylene blue counterstain for 15 sec.

Loeffler's methylene blue
Saturated solution of methylene blue in alcohol
30 ml.
KOH (0·01 per cent. in water)   100 ml.

8 Wash in water, blot and dry in air.

Since tap water may contain saprophytic acid-fast mycobacteria it should not be used to make up staining reagents for use in the Ziehl-Neelsen method; similarly washing of the preparation at all stages during staining should be with water known to be free from such saprophytic species.

ALBERT'S STAINING METHOD *

Apply solution 1; allow to act for 3-5 min.

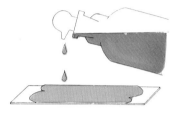

Solution 1

Toluidine blue	0·15 g.
Malachite green	0·2 g.

These are dissolved in 2 ml. of 95 per cent. alcohol and added to 100 ml. of distilled water containing 1 ml. of glacial acetic acid. Ready for use after standing for 24 hr and being filtered.

2 Wash in water; BLOT DRY.

3 Apply solution 2; allow to act for 1 min.

Solution 2

Iodine	2 g.
Potassium iodide	3 g.
Distilled water	300 ml.

4 Wash and blot dry.

* As modified by Laybourn, R. L. (1924): *J. Amer. med. Ass.*, **83**, 121.

CHAPTER 2

PHYSIOLOGY AND CULTIVATION

BACTERIA, as living cells, are subject to a variety of environmental influences in regard to their growth and survival. The following account of bacterial physiology is synoptic and details should be sought in a standard textbook.

Nutrition. The basic requirements of bacteria are very similar to those of higher forms of life; (1) an energy source, (2) a carbon source and (3) a nitrogen source for the synthesis of proteins and nucleic acids and a supply of many inorganic salts and of vitamins and other accessory growth factors. This similarity is not surprising since chemical analysis of bacteria has revealed that their composition is also very similar to higher animals and that the amino acids, nucleotides and fatty acids occurring in bacteria are often identical with these in higher organisms.

However, bacteria vary in their ability to produce these complex organic substances from simple materials; many non-parasitic bacteria can utilise CO_2 as a sole source of carbon and obtain energy for synthetic processes from sunlight (*photo-autotrophic* bacteria) or by oxidation of inorganic material (*chemo-autotrophic* bacteria).

Parasitic species, however, cannot use such simple sources of carbon or energy and must be provided with organic nutrients such as carbohydrates, amino acids, etc. (*heterotrophic* bacteria); in general it can be stated that the more strictly parasitic an organism has become, the more *exacting* it is in regard to its diet. Viruses, as the extreme to the autotrophic bacteria, are unable to synthesise their requirements and are dependent on the host cell for their 'diet'.

Respiration. The majority of bacteria are tolerant in regard to the presence or absence of free oxygen in their environment and grow under a wide range of oxygen tensions—*facultative anaerobes*. Some species will grow only if oxygen is freely available—*obligate aerobes*, e.g., tubercle bacilli, while others cannot grow unless all traces of oxygen have been removed from their environment—*obligate anaerobes*, e.g., tetanus bacilli; indeed many obligate anaerobes not only refuse to grow but are killed if exposed to free oxygen, probably since they then produce peroxides which are rapidly lethal in the absence of the enzyme catalase which is not produced by obligate anaerobes.

Lastly, there are a few bacteria which are *microaerophilic*, i.e., grow more rapidly and luxuriantly in the presence of *traces* of free oxygen.

Temperature. Each bacterial species has a temperature range within which growth will take place and somewhere in this 'minimum-maximum' range lies the 'optimum' temperature at which a particular species grows best; the optimum temperature is usually that of the natural habitat of the bacteria, e.g., approximately 37°C for species parasitic on man and warm-blooded animals. *Mesophilic* bacteria are those that grow best between 25° and 40°C and this group includes all species parasitic on man and some saprophytic bacteria occurring in soil and water; *psychrophilic* bacteria grow best below 20°C and will grow although slowly at temperatures obtained in refrigerators. They are non-pathogenic but have an obvious potential for spoiling refrigerated foods.

A third group are designated *thermophilic* since they flourish at temperatures between 55° and 80°C; they too are non-pathogenic. It should be appreciated that pathogenic and other bacteria although mesophilic are not destroyed by being held at refrigerator temperatures; however, exposure to temperatures significantly higher than the maximum of the growth range is lethal to any species. Thus destruction of bacteria is efficiently carried out by heat; under moist conditions, death is due to coagulative denaturation of the bacterial proteins and exposure to dry heat effects death by oxidation and charring. The higher the temperature employed the shorter the time required for sterilisation but other factors are involved in the amount of heat required to destroy bacteria—they are more easily killed in the presence of acids or alkalies and less readily killed if proteins or other organic materials are present.

The Thermal Death Point (TDP) of any species is defined as the lowest temperature above the maximum at which growth occurs at which the species is killed in a given period of time, e.g., 10 min. For most non-sporing mesophilic species the TDP lies between 50° and 65°C on exposure to moist heat and between 100° and 120°C for most sporing species with the saprophytic *Bacillus stearothermophilus* as the extreme example in requiring 121°C for 10 to 30 min. for destruction of its spores.

Hydrogen-ion concentration. There is wide variation in the tolerance range of pathogenic bacteria; the cholera vibrio for example prefers a pH of 8 and is intolerant of acid environments. On the other hand some organisms are *acidophilic*, e.g., lactobacilli, and flourish in a pH of less than 4.

The majority of commensal and pathogenic bacteria grow best under slightly alkaline conditions, usually 7·2 to 7·6.

Influence of moisture. More than 80 per cent. by weight of the bacterial cell consists of water and as in higher organisms, moisture is essential for growth. Drying under natural conditions is tolerated differently by various species, e.g., tubercle bacilli and staphylococci may survive for weeks or months whereas gonococci die within a very few hours of leaving the human host.

Spores are very resistant to desiccation as indicated by the survival of anthrax spores for more than 50 years when dried on to threads.

By combining *rapid* freezing and drying techniques (lyophilisation) however, even gonococci can be maintained in sealed ampoules and this freeze-drying process is used for the laboratory preservation of cultures.

Influence of light and other radiations. All parasitic bacteria grow and survive best in darkness; ultra-violet rays are rapidly lethal whether derived from an artificial source or naturally, either as direct sunlight or as 'sky-shine', i.e., diffuse sunlight.

CULTIVATION OF BACTERIA

Media. In the isolation of bacteria we cater for their physiological requirements by providing nutrient media dispensed in stoppered test-tubes or screw-cap bottles if fluid or in Petri dishes if solid. Solidification of any fluid medium is easily effected by adding a small amount of agar. *Nutrient broth* is the basis of many media used in diagnostic laboratories and this is a watery solution of peptone and meat extract. Peptone is a mixture of polypeptides and amino acids serving as sources of nitrogen, carbon and energy and is obtained by digesting meat with trypsin or other proteolytic enzymes; meat extract, comprising the water-soluble components of meat, supplements peptone by providing a variety of mineral salts and growth factors. Nutrient broth or nutrient agar have their nutritive value enhanced by addition of blood thus providing a medium adequate for many exacting organisms, e.g., the whooping-cough bacillus.

Plating out. Isolation of pure cultures is essential for the identification of bacteria and this is normally achieved by plating out on a solid culture medium in a Petri dish; a well is spread over a quarter of the surface using a wire loop loaded with fluid culture, pus or using a swab previously inoculated from the infected patient. The loop is then sterilised in a Bunsen flame and after cooling is recharged from the well and drawn in two or three parallel lines across a fresh part of the medium, this procedure is repeated using as inoculum the most distal part of the immediately preceding strokes (page 20). Thus the inoculum density is successively reduced so that ultimately, isolated organisms are deposited and on incubation these give rise to individual colonies.

Each separate colony is a *pure culture* of a single kind of bacterium since it consists exclusively of the progeny of a single bacterial cell.

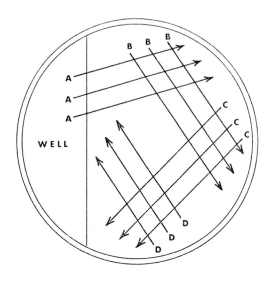

Normal method of plating out. A reducing inoculum from the well area through the series of strokes is ensured by sterilising the inoculating loop at each stage.

STANDARD METHOD
OF PLATING OUT

Plating out on selective media. Such media can be more heavily inoculated and there is no need to sterilise the inoculating loop between each series of strokes.

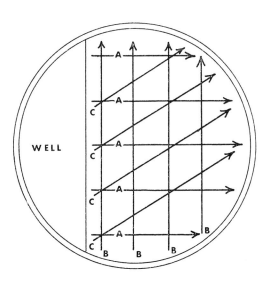

PLATING OUT METHOD
FOR SELECTIVE MEDIA

In circumstances where it is known that mixtures of bacterial species will be present, e.g., in a specimen of faeces, it is essential to employ selective media designed to inhibit the growth of commensal organisms thus allowing any pathogenic species (which may be scanty) to obtain more nutrient; in the case of faecal specimens this is effected by using a desoxycholate citrate agar (DCA) plate; such selective media can be more heavily inoculated and there is no need to flame the loop between successive strokes (page 20). Similarly in such specimens, enrichment media can be used in parallel with the primary selective media and ultimate subinoculation to the latter; enrichment media such as selenite broth and tetrathionate broth definitely improve the isolation rate of intestinal pathogens compared with the use of DCA alone. Having provided adequate nutrients, most bacteria will grow in the natural atmosphere but obligate anaerobes must be grown in the absence of free oxygen.

Incubation. Inoculated media are maintained at the optimum temperature (usually 37°C) by use of a thermostatically controlled incubator or warm-room which is dark and thus also excludes harmful ultra-violet radiations. If incubation is prolonged then the mouths of test-tubes must be sealed with tight-fitting rubber caps or alternatively the medium can be dispensed in screw-cap bottles.

ANAEROBIC CULTIVATION

Many kinds of apparatus have been used for obtaining anaerobiosis in sealed containers and three methods are briefly described.

1. *Electric jar*

A modern type of McIntosh and Fildes jar is shown on pages 22 and 23. There are two needle valves for introduction and extraction of gases, two terminals externally to connect with the palladinised asbestos heater suspended in the jar and the lid is held in place with a screw-clamp handle. A gas-tight seal between the lid and rim of the jar is ensured by smearing the contacting surfaces with a high pressure grease such as 'Apiezon'.

Some earlier models of the McIntosh and Fildes anaerobic jar had a glass body but the potential danger of explosion and injury from flying glass precludes their use nowadays.

Setting up of the jar. Cultures are first placed in the jar and the lid replaced and secured by the screw clamp. One of the needle valves is opened and attached through a gauge to a vacuum pump and the jar exhausted to a negative pressure of 28 to 30 cm. of mercury. The needle valve is then closed and the pump disconnected; a football bladder filled with hydrogen from a cylinder is then connected to the gas inlet and the valve opened to allow the gas to replace the partial vacuum. The terminals on the lid of the jar are connected to a 12-volt electrical supply through a mains transformer and the current switched on.

C

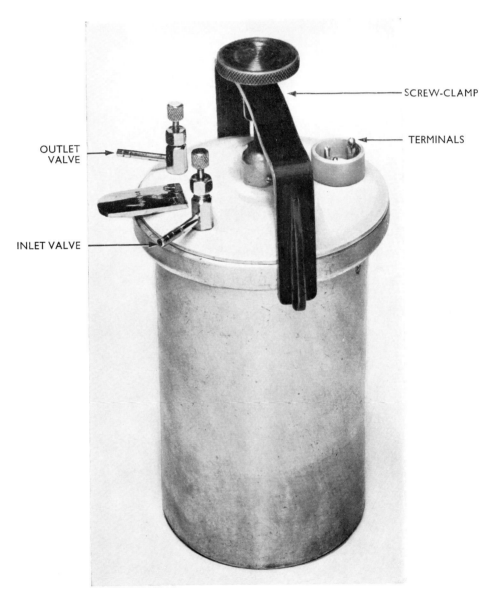

OUTLET
VALVE

INLET VALVE

SCREW-CLAMP

TERMINALS

Anaerobic jar with lid *in situ*.

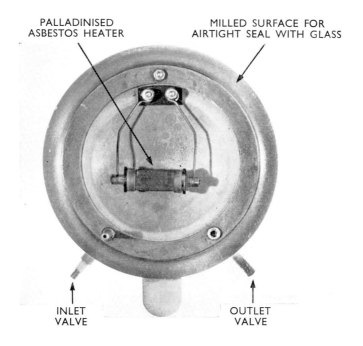

PALLADINISED
ASBESTOS HEATER

MILLED SURFACE FOR
AIRTIGHT SEAL WITH GLASS

INLET
VALVE

OUTLET
VALVE

Under aspect of lid of anaerobic jar.

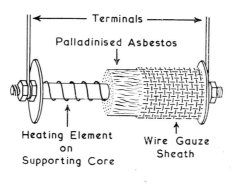

Terminals

Palladinised Asbestos

Heating Element
on
Supporting Core

Wire Gauze
Sheath

CUT-AWAY SECTION OF
PALLADINISED ASBESTOS HEATER

The first effect of heating the coil is that the gases expand and then as catalysis occurs more hydrogen flows into the jar. Hydrogen should be allowed to flow for half an hour with the heater turned on for 10 min. at the beginning and end of this period.

2. *Cold catalyst method*

An equally efficient system which does not require electric operation has been introduced and is available commercially; this type of jar employs a room-temperature or cold catalyst consisting of alumina pellets coated with palladium contained in a gauze sachet which is suspended from the lid of the jar. Heating is not necessary with such cold catalysts so that after the inoculated plates are placed in the jar and the lid firmly fixed a negative pressure of 30 cm. of mercury is drawn in the usual way and the jar is then filled with hydrogen and left attached to the hydrogen source for 15 min. before being incubated.

Cold catalysts are inactivated by moisture and it is essential that the capsule should be completely dry before use; they should be dried in a hot-air oven and then stored in a dry place.

3. *GasPak system**

The GasPak envelope incorporates a system which allows the generation of hydrogen and carbon dioxide simply by adding 10 ml. of water to the contents. Although a specially designed poly-carbonate jar is available, ordinary anaerobic jars can also be used provided that the gas inlets and outlets are firmly sealed. The material to be incubated is placed in the jar in the normal way. A GasPak envelope is opened to a predetermined level and after the water has been added, the envelope is placed immediately and in an upright and unfolded state into the jar and the lid sealed. The presence of a cold catalyst in the jar allows the hydrogen to combine with the oxygen within the jar to give strictly anaerobic conditions. The merits of this recently introduced system are that it obviates the need for drawing a vacuum and eliminates the use of hydrogen cylinders and the fire hazards associated with these. The speed, simplicity and reliability of the method outweigh its cost and extensive trials with the strictest of anaerobic bacteria have proven its efficacy.

An indicator of anaerobiosis must be included in each jar during incubation; this is readily obtained by including a cotton-wool stoppered test-tube containing equal volumes of the following materials:

1. 0·1 N NaOH 6 ml., water to 100 ml.
2. 0·5 per cent. watery methylene blue 3 ml., water to 100 ml.
3. Glucose 6 g., water to 100 ml. and a small crystal of thymol.

After aliquots of these solutions have been mixed in the test-tube the mixture is boiled until it is in the reduced colourless state and then the test-tube is immediately placed in the anaerobic jar. If anaerobic conditions are preserved

* GasPak kits are available commercially from Baltimore Biological Laboratories, Becton Dickinson U.K. Ltd., York House, Empire Way, Wembley, Middlesex HA9 0PS.

during incubation the indicator solution will remain colourless and any defect in the equipment, e.g., a leaking junction between lid and jar body is indicated by the leuco-methylene blue solution regaining its natural colour.

Any medium can thus be inoculated and incubated anaerobically. However, there are many occasions on which one wishes to ascertain whether or not anaerobic bacteria are present and for this purpose Robertson's cooked-meat medium can be used freely and effectively without recourse to anaerobic jars. This medium consists of cooked, minced, sterile muscle and contains reducing substances which maintain anaerobic conditions in the depths of this fluid medium.

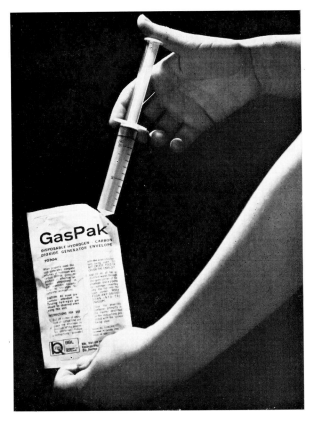

The simplicity of the 'GasPak' system of generating hydrogen and carbon dioxide is shown here; the foil envelope is opened by peeling back the corner to a printed line and 10 ml. of water is injected. The envelope is immediately placed upright in the anaerobic jar, the lid replaced and screwed down and the jar is then ready for incubation.

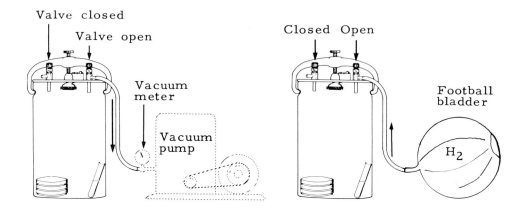

1. EVACUATION

2. REPLACEMENT
WITH HYDROGEN

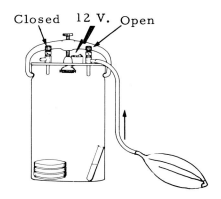

3. CATALYSIS
(Residual O_2 combines with H_2)

Setting up an electric anaerobic jar.

CHAPTER 3

ANTIGENS AND ANTIBODIES

THE study of reactions between antigens and antibodies is termed serology.

AN ANTIGEN is *any* substance which, when introduced *parenterally* into an animal's body, stimulates the formation of antibody and that, when mixed with the latter, reacts *specifically* with it in some *observable* way.

Bacteria and bacterial products are antigenic and in common with other antigens they are usually proteins although some polysaccharides, e.g., pneumococcal capsular material, are efficient antigens.

The *specificity* in reaction of antigens and antibodies is invaluable in diagnostic bacteriology; thus, if a bacterial culture has been identified as belonging to a genus whose members are morphologically and biochemically identical then by using prepared antisera against the several members of the genus we can determine that the 'unknown' culture is one particular serological entity. Likewise the serum from a patient can be tested for the presence of antibodies to particular organisms in tests with standard laboratory cultures.

AN ANTIBODY is a globulin that appears in the blood serum and tissue fluids of an animal in response to the introduction of an antigen and which, when mixed with that antigen, reacts specifically with it in some observable way.

Antibodies are serum proteins, mainly *gamma globulins*, which have been modified so that they react specifically with the antigen which stimulated their production; antibodies appear to be produced by the non-granular monocytic cells, particularly the plasma cells.

The different serological reactions that can be demonstrated with an antiserum are largely determined by the physical state of the antigen employed in the test; thus we speak of:—

Precipitation. When the bacterial antigen is present in colloidal solution and is carefully layered over its specific antiserum, precipitation occurs at the interface (page 29). Alternatively, by incorporating the antiserum in a semi-solid medium such as agar and then introducing the soluble antigen, various bands of precipitate may be noted since different antigens diffuse at different rates; thus one can determine the number of components in a mixture of antigens.

Agglutination. The antigen consists of a suspension of intact bacteria which clump together in the presence of specific antiserum and these clumps aggregate and settle as a deposit with clearing of the supernatant. Such reactions can be made quantitative by employing serial dilutions of the antiserum to each of which is added a constant volume of antigen; thus, the agglutinating titre of a serum is stated as the highest dilution with clearly visible agglutination (page 29).

In diagnostic laboratories, precipitation tests are most frequently employed in determining the group-specificity of β-haemolytic streptococci; the method of extracting the group-specific carbohydrate antigen is shown on p. 51. In this example the left-hand tube contained group-A antiserum and that on the right, group-C antiserum; equal volumes of antigenic extract were then carefully superimposed on each antiserum. Within 5 min. the tube on the left showed heavy precipitation at the interface whilst that on the right remained unaltered. Therefore, the strain belonged to group A, i.e., *Strept. pyogenes*; another method of rapidly identifying these strains of β-haemolytic streptococci which are pathogenic to man is shown on p. 176.

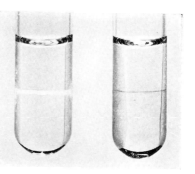

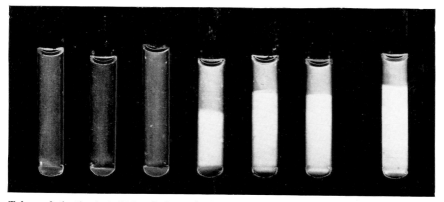

Tube agglutination test. Tubes 1–6 contain doubling dilutions of a patient's serum (30–960) and tube 7 is a control tube without serum which shows that the bacterial suspension is not autoagglutinable. All seven tubes had an equal volume of bacterial suspension (*S. paratyphi B*) added and were then incubated in a water-bath at 37°C for 4 hr. The tubes were then removed from the bath and left at room temperature for 2 hr.
Deposition of the agglutinated bacteria, with complete clearing of the supernatant, has occurred in tubes 1–3; the agglutinating titre of the serum was therefore 120.
In tubes 4–6, and in the control tube, agglutination has not occurred (see p. 165 for details of the technique).

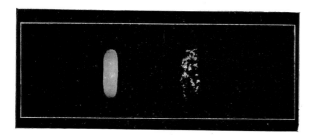

Slide agglutination tests. These are employed as a preliminary step in the identification of many organisms. In this instance loopfuls of a saline suspension of a bacterial culture (? salmonella) were placed separately on the slide; a loopful of group-A salmonella antiserum was added on the left and, on the right, a loopful of group-B salmonella antiserum was similarly mixed with the bacterial suspension. The slide was rocked gently and within 30 sec. agglutination was obvious in the mixture on the right; on the left the bacterial suspension was unaffected and remained homogeneous. Subsequent detailed examination revealed that the organism was *S. typhimurium*. The results of slide agglutination tests must always be confirmed by tube agglutination tests.

Opsonisation. By estimating the number of bacteria ingested by phagocytes in the presence of a patient's serum in comparison with the number phago-cytosed in a non-immune serum an index of opsonic activity can be calculated, e.g., if on counting the bacteria within 50 phagocytes in the presence of a patient's serum the average per phagocyte is 10 whereas in non-immune serum only 2 bacteria per phagocyte are noted, the opsonic index of the patient's serum$=5$. Such tests are technically difficult to standardise and statistically unreliable and are not now in routine use.

Neutralisation. If a bacterium or its toxin is lethal for a laboratory animal then the co-existence of its antibody or antitoxin should be protective to the animal.

In determining whether diphtheria bacilli isolated from a patient are toxin-producing (=virulent), two biologically equivalent guinea-pigs are injected with aliquots of a culture of the bacilli *but* one of the animals will previously have been protected with diphtheria antitoxin. If the strain of diphtheria bacillus is virulent then the protected animal will remain healthy and the unprotected guinea-pig will show characteristic pathological changes; the absence of pathological changes in the latter animal indicates that the organism is avirulent (or that it is not a diphtheria bacillus).

Obviously, and as in all other antigen-antibody reactions, neutralisation tests can also be used to detect antibody if one uses a known antigen; similarly this and other serological reactions can be rendered quantitative by using a series of animals or test-tubes and maintaining a constant concentration of one reagent whilst varying the concentration of the other.

Bacteriolysis. Intact bacterial cells are disrupted in the presence of specific antibody and serum complement.

By themselves, neither antibody nor complement can effect dissolution of bacteria.

Complement is a group of proteins present in the *normal* (non-immune) serum of many animals which deteriorates rapidly *in vitro* unless kept in the frozen state and can be inactivated by heating serum to 55°C for 30 min., which treatment does *not* affect the antibody content.

Haemolysis of red blood cells (RBC) by a haemolytic antiserum is analogous to bacteriolysis, thus:—

RBC+haemolytic serum+complement———→haemolysis.

RBC+haemolytic serum (heated to 55°C/30 min.)———→No haemolysis.

RBC+complement———→No haemolysis.

Thus we have, by suspending red cells in *heated* haemolytic serum, a 'sensitised' or indicator system which when added to another mixture, will either remain unaltered if complement is absent from that mixture or will show a greater or lesser degree of haemolysis if complement is present.

Complement-fixation. This technique depends on two separate reactions. Firstly, antigen and antibody (one of which is unknown) are mixed with a predetermined amount of fresh complement—the antibody-containing serum having been heated to destroy its natural complement. If antigen and antibody are specific to each other then their union will use up or 'fix' the fresh complement; this 'fixation' can only be determined by adding a second 'sensitised' system of RBC suspended in their heated specific haemolytic antiserum. The absence of free complement allows the red cells to remain intact; if, however, the primary antigen and antibody are unrelated, they will not unite and the complement will remain free and available to complete the sensitised system, when it is added, with resultant lysis of the red cells.

In complement-fixation tests, therefore, lysis of the red cells in the sensitised system indicates that the patient's serum in the primary reaction did *not* contain antibody for the particular antigen employed; conversely, the absence of lysis shows that the patient's serum contained specific antibody for the antigen, i.e., the patient's serum was 'positive'.

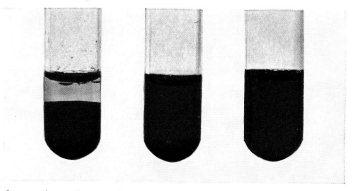

The colour plate shows the result of a *quantitative* complement-fixation test performed with a positive serum, i.e., one which contained antibody specific for the test antigen.

Equal and constant volumes of fresh complement and stock antigen are in each tube with doubling dilutions of patient's serum which had been heated at 55°C for 30 min. to destroy its natural complement; the dilution in the left tube is 1 in 5, in the next tube 1 in 10 and in the tube on the right the serum dilution is 1 in 20.

After primary incubation for 1 hr. at 37°C one volume of sensitised RBC was added to each tube, the contents thoroughly mixed and the tubes replaced in the water bath at 37°C for 30 min.

In the first tube no haemolysis has occurred since there was sufficient antibody to bind all of the complement and none was available to complete the indicator system comprising the RBC with its heated specific haemolytic antiserum.

In the middle tube, although most of the complement was fixed by the primary antigen-antibody reaction the residue allowed partial haemolysis of the sensitised RBC.

31

In the third tube, although antigen and antibody united, there was insufficient antibody to fix most of the complement and enough of the latter was left unbound to effect complete lysis of the sensitised RBC.

If this test had been performed with a patient's serum which was negative, i.e., did not possess antibody specific for the test antigen the result in all three tubes would have been identical and similar to that of the third tube.

BACTERIA

PRIMARY CLASSIFICATION

A. LOWER (EUBACTERIA)

COCCI   BACILLI

VIBRIOS   SPIRILLA

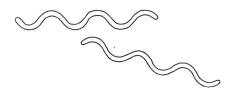

SPIROCHAETES

B. HIGHER (ACTINOMYCETALES)

CHAPTER 4

CLASSIFICATION: DIAGNOSTIC IDENTIFICATION OF BACTERIA

LOWER BACTERIA (eubacteria) are unicellular, never form a mycelium and each cell is biologically independent; they are more numerous and much more important as human pathogens than higher bacteria.

HIGHER BACTERIA are filamentous, often branching and forming a mycelium; frequently cells are interdependent, some being specialised for reproduction.

A broad classification of the lower bacteria is based on cell shape (page 32).

Cocci—spherical or spheroidal.

Bacilli—relatively straight rods.

Vibrios—definitely curved rods.

Spirilla—spiralled, non-flexuous rods.

Spirochaetes—thin, spiralled, flexuous filaments.

Further subdivision of these main categories is essential and that of cocci is shown below. This latter classification is based on the Gram-reaction and the arrangement of cells to each other as determined by *the relationship of consecutive planes of division as daughter cells are produced.*

Thus **Diplococci** are Gram-positive cells adherent in pairs or in very short chains; cells in each pair are slightly elongate in the axis of the pair (line-ahead formation).

Streptococci are Gram-positive cells adhering in chains since successive cell divisions occur in the same plane.

Staphylococci are also Gram-positive but adhere in irregular clusters since consecutive planes of division are haphazard.

CLASSIFICATION OF COCCI

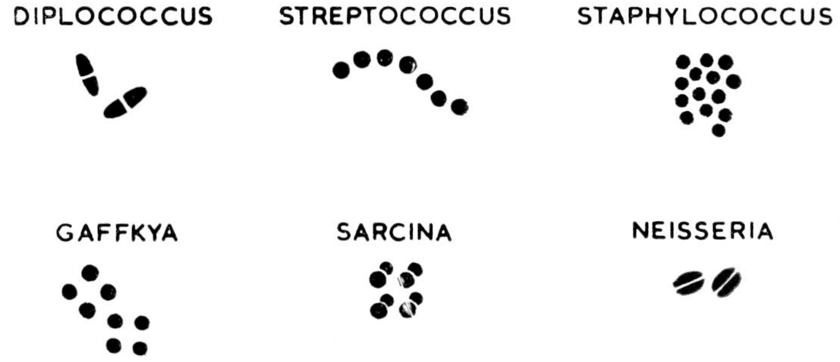

DIPLOCOCCUS   STREPTOCOCCUS   STAPHYLOCOCCUS

GAFFKYA   SARCINA   NEISSERIA

CLASSIFICATION OF BACILLI

A. ACID-FAST

B. GRAM POSITIVE

SPORING   Non-SPORING

includes the following genera[*]
CORYNEBACTERIUM
ERYSIPELOTHRIX
LACTOBACILLUS
LISTERIA

Aerobic

Anaerobic

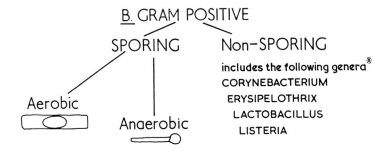

C. GRAM NEGATIVE

includes numerous genera[*]
e.g. PSEUDOMONAS
SALMONELLA
SHIGELLA
PROTEUS
PASTEURELLA
BRUCELLA etc.

[*] can be recognised and differentiated only by further study

Neisseriae stain Gram-negatively and mainly adhere in pairs; the cells are elongate at right angles to the axis of each pair (line-abreast formation).

Gaffkyae appear as flat plates of four cells due to consecutive planes of division being at right angles to each other; Gram-positive.

Sarcinae are seen as cubes or packets of eight cells due to division occurring successively in three planes at right angles; Gram-positive.

BACILLI cannot be subdivided in such a straightforward manner and various characters must be taken into account even for elementary differentiation; the classification outlined on page 34 is made on the response to Gram and Ziehl-Neelsen staining techniques. Further subdivision of Gram-positive bacilli depends on whether or not they can form spores and spore-forming members can be further characterised by their ability or inability to grow aerobically. Detailed recognition of genera within the non-sporing Gram-positive and within the Gram-negative bacilli requires further investigation not only of morphological but of cultural and biochemical characteristics.

VIBRIOS and SPIRILLA are Gram-negative and mostly motile having polar flagella; differentiation of species is by cultural, biochemical and serological methods.

PATHOGENIC SPIROCHAETES may be classified into three genera (below); **Borrelia,** larger than other pathogenic members, can be stained and seen by ordinary methods, Gram-negative; **Treponema**—much finer and possess more coils than Borrelia, only demonstrable by dark-ground microscopy or if stained by a silver impregnation method; **Leptospira**—even finer and possess more numerous coils which are so close to each other as to be barely discernible, one or both ends are recurved on the body of the organism.

CLASSIFICATION OF SPIROCHAETES

BORRELIA

TREPONEMA

LEPTOSPIRA

DIAGNOSTIC IDENTIFICATION OF BACTERIA

The following brief summaries of the steps taken to identify an organism isolated from pathological material outline the procedures followed in the laboratory; in the following chapters each genus is dealt with under these headings.

MICROSCOPY is invaluable as a simple swift method of placing bacteria in their appropriate group according to their staining reactions and morphology; rarely, however, can this essentially preliminary step be sufficient in itself for diagnostic identification. Noteworthy exceptions are the presence of acid- and alcohol-fast bacilli in sputum (tubercle bacilli) and the presence of intracellular Gram-negative diplococci in cerebrospinal fluid (meningococci) or in urethral discharge (gonococci). In one or two instances, e.g., leprosy and Vincent's infection, microscopy is at present the only available diagnostic method. Microscopic characteristics which should be noted routinely are size, shape, arrangement of cells to each other, Gram-reactivity, presence or absence of motility, capsules and/or spores.

CULTURAL REQUIREMENTS AND APPEARANCES are a prerequisite in separating many species which are microscopically identical. Colonial characteristics which should be noted include the size of colony, its shape in plan view and elevation, opacity and translucence, pigmentation—either natural as in staphylococci or resulting from indicator change in differential media, e.g., *Escherichia coli* on MacConkey's medium. The consistency of the colony and any changes in the colour or consistency of the underlying medium should also be noted. Relevant colonial features are dealt with for each genus but common colonial types are exemplified below.

ELEVATIONS AND EDGES
OF COMMON TYPES OF COLONIES

ELEVATIONS

Flat   Low Convex   High Convex   Plateau

EDGES

Entire   Irregular   Crenated

BIOCHEMICAL REACTIONS OF CULTURES are constantly employed to differentiate pathogenic species within a genus from each other and from commensal members of the genus.

The varying ability of bacteria to ferment carbohydrates with acid production, which may or may not be accompanied by the evolution of gases, is widely used and is particularly valuable in the differentiation of Gram-negative intestinal bacteria. Fermentation of a carbohydrate is tested by growing the organisms in a fluid medium containing usually 1 per cent. of the sugar and an indicator which reacts tinctorially if acid is produced; an inverted miniature test tube (Durham tube) is immersed in the medium and gas formation is revealed by the collection of bubbles at its apex.

Other enzymatic activities may be tested, e.g., indole and H$_2$S production from peptone, coagulase activity of pathogenic staphylococci, etc.

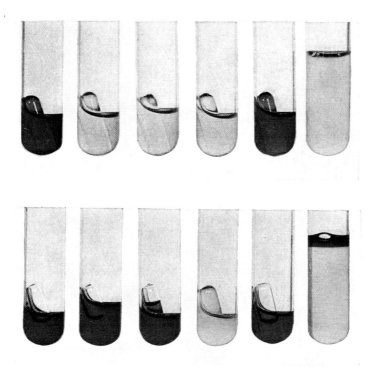

Biochemical differentiation of microscopically similar organisms.

1 per cent. solutions of glucose, lactose, dulcitol, sucrose and mannitol in peptone water are contained respectively in tubes 1–5; the sixth tube contains plain peptone water.

The upper row of tubes was inoculated with an organism (*Shigella sonnei*) which ferments only glucose and mannitol without producing gas. Indole is not produced in the peptone water culture.

The lower row of tubes was inoculated with an organism (*Escherichia coli*) which, with gas production, utilised all substrates except sucrose and also produced indole. The production of indole from peptone is readily tested by adding 1 ml. of ether to the peptone water culture which is then shaken vigorously, and, after allowing the tube to stand for 2 min., 0·5 ml. of Ehrlich's rosindole reagent is added.

37

D

SEROLOGICAL REACTIONS. When an animal host is infected naturally or experimentally with pathogenic bacteria, recovery is often associated with an enhanced ability to withstand subsequent challenge by the *same* bacteria; this increased resistance is in many cases attributable to the appearance of antibody in the blood serum and tissues of the host. Such antibodies have demonstrable and specific reactivity *in vitro* with the antigens which stimulate their formation. An unknown bacterium can thus be identified by demonstrating its reaction with one of several standard antisera; similarly by using laboratory cultures of fully identified organisms one can demonstrate the presence of specific antibodies in the serum of an individual suffering from or recovered from many infections.

ANIMAL INOCULATION is occasionally used in the identification of certain pathogenic bacteria, e.g., tubercle bacilli and leptospirae, which are virulent to and produce characteristic lesions in guinea-pigs or other laboratory animals. In general such use of animals is restricted as far as possible not only for humanitarian reasons but because of expense.

SECTION II

SYSTEMATIC BACTERIOLOGY

TABLES OF DIFFERENTIAL CHARACTERISTICS

The various tables included in this section of the text are intended to exemplify the use of biochemical and other tests in the recognition of groups or species within various genera; **it is not the authors' intention that these should be memorised.**

The following key can be used for all of these tables:—

Fermentation reactions:

$\perp$=Acid production without gas formation.

$+$=Acid and gas produced.

$-$=No fermentation.

( )=Delayed reaction.

Other reactions:

Gelatin liquefaction

$+$=liquefaction occurs.

$-$=no liquefaction.

Indole production

$+$=indole produced.

$-$=no indole produced.

Motility

$+$=motile.

$-$=non-motile.

Growth, *e.g., on MacConkey's medium*

$+$=growth occurs.

$-$=no growth.

The abbreviation 'V' in any table indicates that strains within a group are variable, e.g., some *Shigella flexneri* strains produce indole and other strains do not.

CHAPTER 5

STAPHYLOCOCCI

STAPHYLOCOCCI are ubiquitous and can be isolated from air, dust, foodstuffs and from human beings and animals. Some are members of the natural flora of skin and mucous surfaces, others are capable of causing infections of varying severity.

Microscopy. Approximately 1μ in diameter, spherical, arranged irregularly in clusters, Gram-positive, non-motile, non-sporing, and non-capsulate.

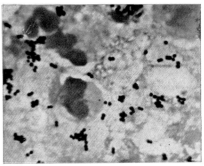

Gram-stained film of pus showing the characteristic arrangement and other features of staphylococci
× 1000.

Cultural appearances. Grow abundantly on all ordinary media, facultatively anaerobic; colonies have an entire edge and convex elevation showing white, golden or lemon pigmentation and 2 to 4 mm. in diameter after 18 to 24 hr incubation at 37°C. Pigmentation is less marked or absent when cultures have been incubated anaerobically but develops when such cultures are exposed to air. Cultivation on milk agar accentuates pigment production but no reliance is now placed on particular pigment production as a guide to pathogenicity.

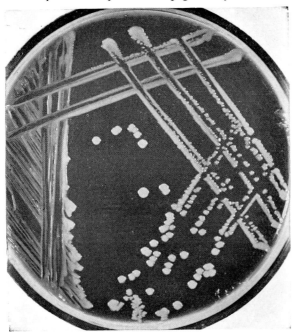

Blood agar plate inoculated from a nasal swab. After 18 hr incubation at 37°C a pure and profuse growth of *Staph. pyogenes* var. *aureus* was evident. These proved to be coagulase +ve indicating that the individual from whom the swab was taken was a carrier of pathogenic staphylococci. The size of these colonies should be contrasted with the much smaller (and non-pigmented) colonies of streptococci. (See pp. 48 and 52).

Biochemical reactions. Staphylococci vary in their ability to ferment sugars and to liquefy gelatin. Earlier attempts at classification on such features have been replaced by testing the coagulase production of cultures.

COAGULASE TEST. Coagulase is a prothrombin-like enzyme produced by more than 96 per cent. of staphylococci isolated from human infections; contrarily, commensal staphylococci from human skin and saprophytic species rarely produce coagulase. This high correlation between coagulase production and pathogenicity provides to date the most convenient and reliable single test for determining the pathogenicity of a given strain.

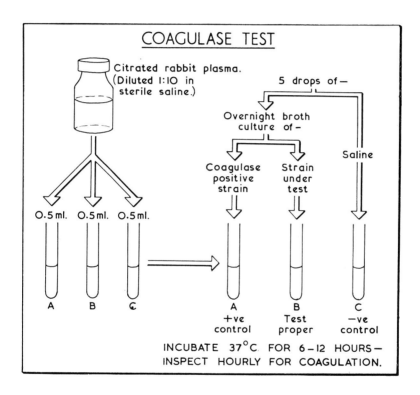

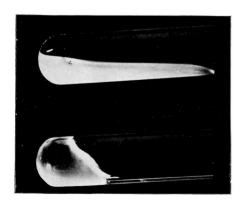

Coagulase tests after incubation at 37°C for 6 hr.
The upper tube shows a negative test; the contents are fluid.
In the lower tube the citrated plasma has been coagulated, indicating a positive result.

Coagulase +ve staphylococci are pathogenic and named *Staph. pyogenes*; such strains usually produce golden-yellow colonies (var. *aureus*), occasionally white colonies (var. *albus*) and rarely lemon pigmented colonies (var. *citreus*).

In contrast, coagulase −ve staphylococci are commensal and termed *Staph. epidermidis* since they commonly occur on the skin.

Serological characters. Although agglutination and precipitation tests can separate pathogenic from non-pathogenic strains and also allow the recognition of several serotypes within the pathogenic group, the information thus obtained in epidemiologic studies of staphylococcal infections has been less satisfactory than that obtained by phage typing.

PHAGE TYPING

Bacteriophages are viruses which are parasitic on bacteria and display highly specific relationships with their host cell; like all viruses they multiply only within the host cell and their size precludes direct visualisation other than by electron microscopy. However, phages are able to lyse many of their bacterial hosts so that by implanting phage preparations on a plate seeded with a culture of the susceptible bacteria one can, after incubation, note activity of the phage macroscopically by the presence of plaques or zones of lysis indicating specific lytic activity of the phage on the bacteria. Thus, by selecting appropriate phages for a given bacterial genus it is possible to distinguish phage types in the latter by noting which of the standard collection of phages are active against a particular strain of the genus.

In an epidemic situation, e.g., post-operative wound infection, phage typing of the staphylococci from the wounds, from the anterior nares of carriers, from air samples and swabbing of fomites, etc., will reveal the source and spread of the epidemic type in the environment and on this evidence measures can be instituted to eliminate the source.

Animal inoculation has no place in diagnostic procedures for staphylococci.

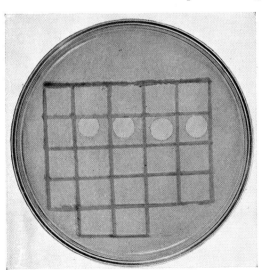

Phage typing of staphylococci

A plate of digest agar has been marked with a grid of 22 squares, each corresponding to the site of a particular phage preparation. The surface of the plate was flooded with a broth culture of a staphylococcus and then allowed to dry. The phage preparations were then individually applied each in 0·01 ml. amounts and when these had dried the plate was incubated overnight at 37°C. On examination confluent lysis was noted in the areas of phage preparations 3B/3C/55/71.

PHAGE TYPING TECHNIQUE

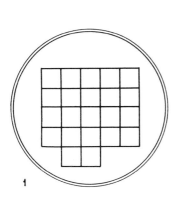

1

GRID PLATE OF DIGEST
AGAR

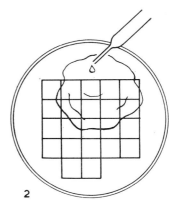

2

SURFACE FLOODED WITH
OVERNIGHT BROTH CUL-
TURE OF STRAIN TO BE
TYPED

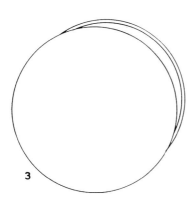

3

LID REPLACED INCOM—
PLETELY. LEFT AT ROOM
TEMPERATURE FOR 1—3 HR.

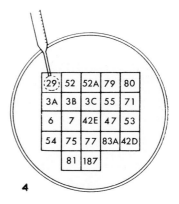

4

0·01 ml. DROP OF EACH
TYPING PHAGE APPLIED.
PLATE ALLOWED TO DRY
— INCUBATED AT 37° C
FOR 12—18 HR.

INFECTIONS CAUSED BY STAPHYLOCOCCI

Although there are occasional reports of serious and sometimes fatal infection due to coagulase -ve staphylococci, *Staph. pyogenes* is incriminated in most instances.

Staphylococcal surface infections range from the simple follicular lesion involving only one hair root through the more severe skin carbuncle and breast abscess to pemphigus neonatorum; staphylococci are also incriminated in some cases of blepharitis, impetigo and ophthalmia neonatorum.

Among the more deeply seated lesions caused by staphylococci acute osteomyelitis, particularly of the long bones, was once among the most dreaded of diseases; in the pre-antibiotic era, even with speedy diagnosis, osteomyelitis frequently had a chronic course involving intermittent surgical removal of bone sequestra with sinus formation and sores which lasted for many years. Nowadays with early diagnosis, isolation of the causal staphylococcus and prompt determination of its antibiogram few cases reach the chronic stage which previously often required limb amputation to save the patient from the slow but inevitable outcome of amyloid disease. Other deep-seated infections caused by staphylococci include broncho-pneumonia, peri-renal abscess and occasionally staphylococcal infection can spread from the original site and become septicaemic with multiple abscesses in many tissues.

Certain strains produce a heat-stable enterotoxin which if ingested gives rise to acute toxic food poisoning which often occurs in epidemic fashion.

The staphylococcus still plays an important role in human disease particularly in hospital-acquired infections; this is frequently in the form of postoperative wound infection or infection of burns. The added danger of such infections is that hospital strains of staphylococci have, with a rapidity and ease perhaps unrivalled by any other genus, been able to adapt themselves to the antibiotic environment so that hospital strains are invariably resistant to many more therapeutic agents than strains encountered in the open community.

A discussion of the preventive measures which should be invoked to reduce, if not eliminate, hospital-acquired infection is beyond the scope of this book, e.g., the control or elimination of carrier state in hospital personnel, the cleaning and subsequent sterilisation of instruments, etc., the ventilation and design of wards, operating theatres and dressing stations and the need for intelligent realisation of the reasons for many procedures rather than performance of blind ritual. But since we have little or no knowledge of the incidence of such infections the impact of such preventive procedures cannot be measured accurately.

Hence there is an urgent need for detailed notification and investigation of each case and epidemic of hospital-acquired infection since with such information the benefits accruing from the introduction of one or other prophylactic regimes could be evaluated fully.

CHAPTER 6

STREPTOCOCCI

STREPTOCOCCI are widely distributed in nature; some are members of the normal human flora and others cause human diseases directly attributable to infection or indirectly by sensitisation to them.

Microscopy. Regardless of the varying cultural, biochemical and serological characteristics of streptococci *members of this genus are microscopically indistinguishable.* Approximately 1μ in diameter, spherical, arranged in chains of varying length, Gram-positive, non-motile, capsulate in certain circumstances and non-sporing.

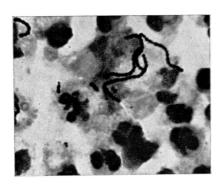

Gram-stained film of pus containing *Streptococcus pyogenes.* Streptococci can be seen in pairs and in chains of varying length; the chain length is no guide to the cultural type of streptococcus. ×1000.

Cultural appearances. Some members of the genus are obligate anaerobes and these are dealt with at the end of this chapter. Streptococci are more fastidious than staphylococci in their cultural requirements; they grow best on blood agar which also allows a primary differentiation of the *aerobic* streptococci. Three responses can be seen surrounding streptococcal colonies in blood agar BUT the colonies themselves are very similar in size and shape—they are only half the diameter (1 to 2 mm.) of staphylococcal colonies after 18 to 24 hr incubation at 37°C and are non-pigmented.

α-Haemolytic streptococci are surrounded by a narrow halo of greenish-grey discoloration in which only a few red cells are lysed when the medium is examined microscopically.

β-Haemolytic streptococci are surrounded by a very much wider zone of complete clearing of the red cells which are entirely disrupted.

γ-Haemolytic (non-haemolytic) streptococci produce no visible alteration in the blood agar.

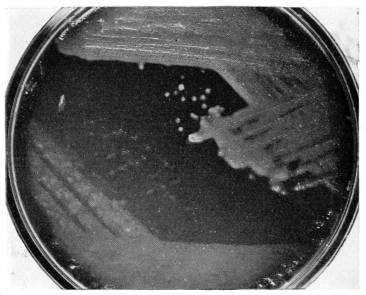

The organisms seen in the upper half of this blood agar plate show β-haemolysis and those in the lower part portray the much less distinct nature of α-haemolysis. The organisms are respectively *Strept. pyogenes* (type 4) and *Strept. viridans*.

These cultural types of streptococci are also clearly differentiated by their pathogenicity and communicability in the human host; α-haemolytic types (*Streptococcus viridans*) are essentially commensal in the healthy upper respiratory tract. γ-Haemolytic types (*Streptococcus faecalis; Enterococcus*) lead a commensal existence in the intestine of man and animals.

In certain circumstances α and γ types can assume a pathogenic role in the human host and such infections are *endogenous* in nature; by contrast, β-haemolytic streptococci account for most of the primary streptococcal infections in man and *exogenous* infection is the rule, i.e., β-haemolytic streptococci spread readily from one host to another in epidemic fashion.

β-Haemolytic Streptococci

Biochemical reactions. Biochemical subdivision has been attempted in an endeavour to establish epidemiologically significant types of β-haemolytic streptococci; such was not satisfactory and serological grouping and typing methods are available and reliable.

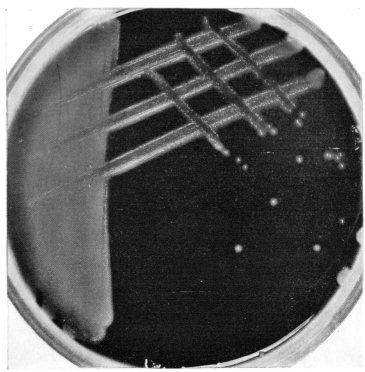

Blood agar plate with a 1 in 500,000 concentration of crystal-violet incorporated. The plate was inoculated from a throat swab taken from a patient
with acute tonsillitis. After 18 hr incubation at 37°C a pure growth of
β-haemolytic streptococci was evident; these proved to belong to Lancefield
group A, type 12. The addition of crystal violet reduces greatly the growth
of many other organisms., e.g., staphylococci.

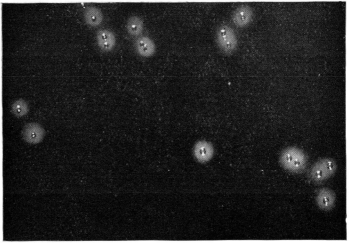

An enlarged view of one area of the above plate reveals the characteristics
of streptococcal colonies which, regardless of the lytic changes effected in
blood agar, are similar for α, β and γ types of streptococci.

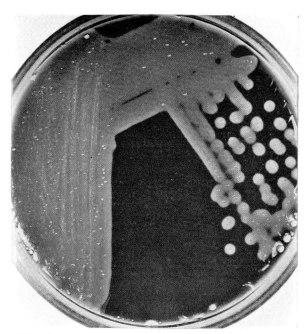

This plate, inoculated from the same throat swab, was incubated **anaerobically.** The enhancement of β-haemolysis under these conditions is obvious.

Serological characters. A broad grouping of β-haemolytic strains is possible by precipitation techniques in which the group-specific carbohydrate (C) antigen extracted from the organism is tested against standard group-specific antisera. The groups A to S (no groups designated I or J) are in general related to different animal hosts and the majority of strains from man belong to group A.

Some Groups of β-haemolytic streptococci

Group	Principal Host
A	Man. All except types 7, 16, 20 and 21 belong to this group.
B	Cattle.
C	Several animals. Man for types 7, 20 and 21.
D	Man.
G	Essentially commensal in man; serotype 16 belongs to this group.
L	Dogs.
M	Pigs.

Since group-A strains (*Streptococcus pyogenes*) are so prevalent in human infections they have been subjected to vigorous serological analysis which now allows the recognition of at least 40 serotypes based on the content of various specific protein (M and T) antigens; strains may possess both, either or neither of these type-specific antigens. Typing of *Strept. pyogenes* is a two-step procedure—firstly, intact cells are tested by slide agglutination with antisera against various T-antigens revealing the presence of one or more T-antigens in the particular strain; the second stage uses acid-extracted M-antigen (the process is identical with that which extracts the group or C-antigen) tested for precipitation with highly type-specific M-antisera.

Strains possessing type-specific M-antigen give matt-surfaced colonies and are virulent compared with the same strain lacking M-antigen; in man, recovery from group-A streptococcal infection is associated with long-lasting immunity against the particular M-serotype of the infecting strain but no protection against the other serotypes within the group. Thus, in theory at least, one individual could suffer at least 40 separate attacks from group-A streptococci.

STREPTOCOCCUS VIRIDANS

Streptococcus viridans has already been noted for its essentially commensal nature in the upper respiratory tract; however, in an individual with a history of rheumatic carditis (usually mitral stenosis) or a congenitally abnormal heart valve (usually a bicuspid aortic valve) *Strept. viridans* is the commonest cause of subacute bacterial endocarditis. This condition was invariably fatal before the advent of antibiotics and even yet carries a 20 to 35 per cent. fatality rate; infection is *endogenous*, the source of the organisms being the mouth in a state of poor dental hygiene. Other organisms are involved in about 5 per cent. of cases of subacute bacterial endocarditis and it is essential that blood culture be performed, and repeatedly, so that the causal organism can be isolated and its sensitivity to antibiotics determined as a guide to rational therapy.

GROUP IDENTIFICATION OF
β- HAEMOLYTIC STREPTOCOCCI.
Lancefield's Method

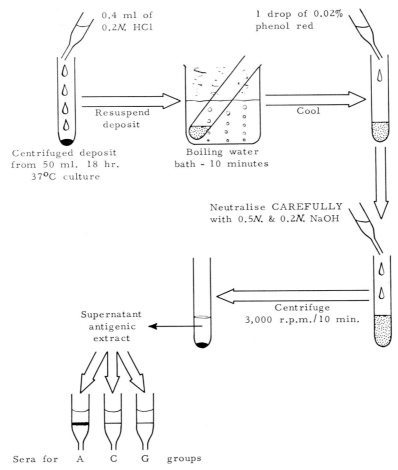

0.4 ml of
0.2*N.* HCl

1 drop of 0.02%
phenol red

Resuspend
deposit

Cool

Centrifuged deposit
from 50 ml. 18 hr.
37°C culture

Boiling water
bath - 10 minutes

Neutralise CAREFULLY
with 0.5*N.* & 0.2*N.* NaOH

Supernatant
antigenic
extract

Centrifuge
3,000 r.p.m./10 min.

Sera for   A   C   G   groups

Allow to stand for 5 minutes ;
precipitation within this time
occurs only with the group
serum specific for the extract.

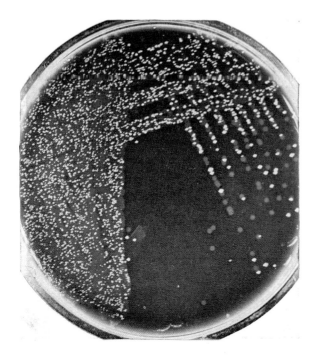

Blood agar plate inoculated from a swab taken from the skin. After overnight incubation at 37°C it can be noted that two species were present, the white pigmented colonies are those of staphylococci which were coagulase —ve and α-haemolytic colonies which were shown to be those of *Strept. viridans*. The much smaller size of the streptococcal colonies can be noted since, even including the surrounding lytic zone, the area occupied by the colonies of *Strept. viridans* just equals the size of the staphylococcal colonies.

Biochemical reactions. These have been examined in an attempt to differentiate species but with little success. The only significant reactions are those that assist in differentiating viridans streptococci from pneumococci (q.v.).

Serological characters. Little is known of these except that *Strept. viridans* does not possess carbohydrate group antigens like β-haemolytic streptococci; their commensal role and the fact that they do not cause disease by exogenous transmission has not encouraged detailed serological studies.

Streptococcus faecalis (*Enterococcus*) can give rise to endogenous urinary tract infections but is a normal inhabitant of the intestine.

Microscopically they are slightly ovoid in shape but no stress can be put on this for differentiation from other streptococci.

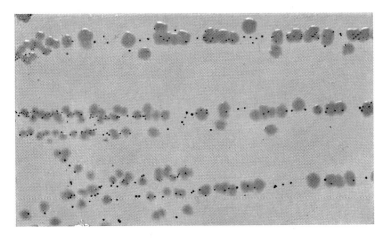

MacConkey's plate showing *Strept. faecalis* colonies as minute and magenta coloured. Their size can be contrasted with those of the larger colonies of *Escherichia coli.* ×3

Biochemical reactions. *Strept. faecalis* can be typed biochemically and in contrast to the other two cultural types of aerobic streptococci is capable of growth in the presence of bile salts, e.g., on MacConkey's medium when they appear as minute (0·5 to 1 mm.) magenta coloured colonies; similarly they will grow in a medium with a high (6·5 per cent.) concentration of NaCl and can withstand exposure to 60°C for 30 min. which kills other streptococci.

Although the majority of faecal streptococci are without effect on blood agar they belong to group D in Lancefield's classification.

Group D streptococci: Biochemical subdivision

Type	Sorbitol	Arabinose	Gelatin Liquefaction	Growth at pH=9·6
Strept. faecalis var. *faecalis* . .	⊥	—	—	+
var. *liquefaciens*[*] .	⊥	—	+	+
var. *zymogenes*[**] .	⊥	—	+	+
Strept. faecium . .	—	⊥	—	+
Strept. durans . .	—	—	—	+
Strept. bovis . .	—	⊥	—	—

[*] No haemolysis on blood agar plate.
[**] β-haemolysis on blood agar plate.

53

E

ANAEROBIC STREPTOCOCCI

These are the most common of all anaerobic cocci, and although they have been found free in nature their normal habitat is the mouth, genito-urinary tract, other mucous surfaces and the skin of healthy persons. One species, *Peptostreptococcus putridus*, seems to be the most prevalent as a potential pathogen and has been investigated in the laboratory more completely than the others.

PEPTOSTREPTOCOCCUS PUTRIDUS

Microscopy. Very similar to the aerobic streptococci but usually smaller in diameter particularly on primary cultivation.

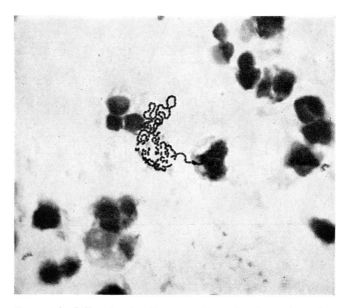

Gram-stained film showing *Pepto. putridus* and polymorphonuclear leucocytes; the film was made from a high vaginal swab from a patient with puerperal sepsis. ×1000.

Cultural appearances. Strictly anaerobic cultivation is required. Ordinary blood agar plates are used but some species other than *Pepto. putridus* will grow more readily if 0·01 per cent. sodium oleate is incorporated in the medium; this constituent does not adversely influence the growth of *Pepto. putridus*. Colonies are only 1 to 2 mm. in diameter after 48 hr incubation at 37°C, smooth in appearance with an entire edge and a low convex elevation. No alteration is noted in the medium.

Biochemical activities. A sulphur compound such as sodium thioglycollate (0·1 per cent.) must be present with the substrate to be tested for fermentation. Glucose, maltose and fructose are fermented with the evolution of foul-smelling gas.

Serological characters. Investigations are at present too incomplete to evaluate the use of serological tests for identification or epidemiological studies.

Animal inoculation. Little is known of their pathogenicity for laboratory animals.

INFECTIONS CAUSED BY STREPT. PYOGENES

It has already been noted that *Strept. viridans* and *Strept. faecalis* lead an essentially commensal existence in the human host but each can give rise to endogenous infection. On the other hand *Strept. pyogenes* is readily spread from cases of infection and carriers to other susceptible individuals; such exogenously acquired infections are particularly common in younger people, especially school children, and include streptococcal sore throat, scarlet fever, otitis media, erysipelas, and some cases of impetigo. *Strept. pyogenes* also infects wounds and burns; many cases of puerperal sepsis were formerly caused by such strains but they are less commonly incriminated nowadays.

Infection with any of the specific serotypes of *Strept. pyogenes* may be followed by acute rheumatic sequelae and this causal relationship has been established not only on clinical grounds but also on the basis of bacteriological and serological studies. The most convincing proof of this relationship, however, was the demonstration of the remarkable prophylactic benefits accruing from long-term administration of suitable antimicrobial agents to known rheumatic subjects and similarly the fact that rheumatic sequelae do not occur if the primary streptococcal infection is adequately treated, i.e., the prompt administration of penicillin in proper dosage for a sufficient period, e.g., one mega unit of depot penicillin given by intramuscular injection will maintain adequate blood-levels for at least 10 days.

Certain serotypes of *Strept. pyogenes*, particularly type 12 strains, may be nephrotoxic; thus acute glomerulo-nephritis may result from infection. It appears that even with rapid and adequate treatment of the primary streptococcal lesion the prevention of glomerulo-nephritis is not guaranteed.

CHAPTER 7

PNEUMOCOCCI

THE pneumococcus (*Strept. pneumoniae*) is a normal inhabitant of the human upper respiratory tract. They can cause pneumonia (usually of the lobar type), para-nasal sinusitis, otitis and meningitis which is usually secondary to some other pneumococcal infection.

Microscopy. 1μ in their long axis, slightly elongated and arranged in pairs when a lanceolate shape may be seen or in very short chains in the line of the long axes, Gram-positive, non-motile, capsulate and non-sporing.

The above description holds for films made from pathological material or in freshly isolated cultures from such material; on continued laboratory cultivation pneumococci are microscopically indistinguishable from streptococci particularly in the disappearance of their capsules.

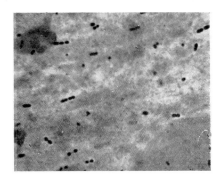

Gram-stained film of sputum from a case of lobar pneumonia; pneumococci appear mainly in pairs. *N.B.* The capsules are not demonstrated by Gram's staining method. ×1000.

Cultural appearances. Facultatively anaerobic; as fastidious as streptococci and produce a zone of α-haemolysis on blood agar hence differentiation from commensal *Strept. viridans* is essential. With both organisms α-haemolysis is accentuated when grown on chocolate blood agar. Colonies are plateau in elevation after 18 to 24 hr incubation at 37°C and on continued incubation autolysis of the centre of the colony is noted—the draughtsman colony.

Biochemical reactions. These are of interest only in so far as they allow differentiation from *Strept. viridans*; the majority of strains ferment inulin as compared with these streptococci but more specific tests are available.

BILE SOLUBILITY TEST. Pneumococci are soluble in bile whereas *Strept. viridans* remain intact. This feature can be tested by adding 1 part of a sterilised 10 per cent. solution of sodium taurocholate in normal saline to 10 parts of a broth culture. Incubation of the mixture for 15 min. at 37°C will reveal whether lysis occurs; it is essential that the broth culture should be adjusted so that its pH is between 7 and 7·5 otherwise chemical precipitation of the bile salt may occur with resultant turbidity.

OPTOCHIN SENSITIVITY. 'Optochin' is an alkaloid allied to quinine and to which pneumococci are extremely sensitive and viridans streptococci are resistant; the test is performed by placing a filter paper disc, impregnated with a 1 in 4000 aqueous solution of 'Optochin', on the surface of a medium which has been sown with the organism.

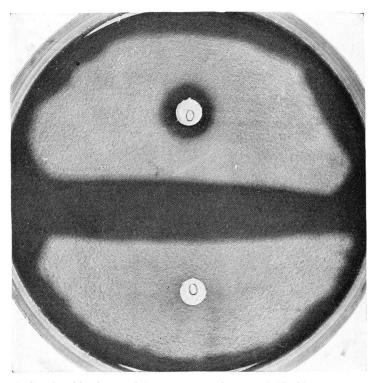

A chocolate blood agar plate was sown on its upper half with a pneumococcal culture and on the lower half with *Strept. viridans*. Filter paper discs, impregnated with a 1 in 4000 solution of 'Optochin', were placed in each area; after 18 hr incubation at 37°C inhibition of the pneumococci contrasts with the uninhibited growth of *Strept. viridans*. The frequent co-existence of potentially pathogenic pneumococci and essentially commensal *Strept. viridans* in specimens from the respiratory tract demands an easy and reliable means of differentiating them. 'Optochin' sensitivity testing is much more reliable than inulin fermentation and more readily performed than bile solubility tests.
The accentuation of α-haemolysis when such species are grown on such a medium is obvious.

Serological characters. At least 77 specific serotypes of pneumococci can be identified; specificity is endowed by the chemically specific capsular polysaccharides. Typing is performed on a slide by mixing a loopful of broth culture or saline suspension from a blood agar plate culture with a loopful of diagnostic antiserum; a cover slip is applied over the mixture and the preparation viewed with an oil-immersion objective under reduced illumination.

In the presence of its specific antiserum the capsule is sharply outlined and appears swollen in comparison with parallel preparations containing heterologous type sera. In the absence of specific antibody the capsule and its margin are quite invisible. In practice, strains are first tested in 9 pooled sera and then in the type-specific sera comprising the pooled serum in which a reaction was obtained; such typing was an essential preliminary in the pre-sulphonamide era when serum therapy was employed in the treatment of pneumococcal lobar pneumonia; nowadays serotyping of pneumococci is only of epidemiological interest.

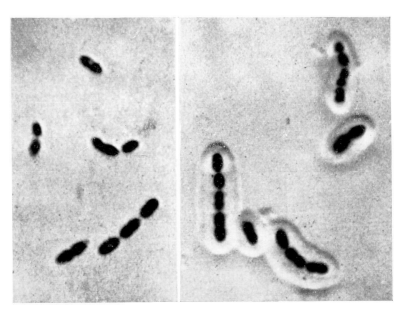

The organism in both preparations was Pneumococcus type 19; in the left-hand plate the organism was in the presence of a *heterologous* antiserum and methylene blue in a wet film. No capsule 'swelling'. In the right-hand plate *homologous* (type 19) antiserum was present and a capsule 'swelling' reaction is obvious. × 3750.

INFECTIONS CAUSED BY PNEUMOCOCCI

Pneumococci are associated quite indelibly, and rightly so, with *lobar pneumonia* in which they are the most common species encountered; however, pneumococci are also commonly involved as secondary pathogens in cases of *bronchopneumonia* where the primary disease is viral in origin, e.g., measles and influenza.

They are also frequently encountered alone or in association with other pyogenic organisms in cases of paranasal sinusitis and acute otitis media; in such cases infection may have been acquired exogenously or alternatively the pneumococci have extended from the naso-pharynx where they have been existing as commensals. Pneumococci frequently spread from the initial focus of infection, e.g., pneumococcal meningitis occurs as a complication of otitis media and lobar pneumonia but some cases of pneumococcal meningitis present as a primary infection. When such patients suffer recurrent attacks diligent investigation frequently reveals that there is a skull defect; such defects may be congenital, e.g., absence of a cribriform plate or post-traumatic. Closure of the defect with a fascial graft gives a high degree of protection against recurrences.

CHAPTER 8

NEISSERIAE

MOST members of this genus are commensal in the upper respiratory tract and occur extra-cellularly; the two pathogenic species (*Neisseria meningitidis*, the meningococcus, and *Neisseria gonorrhoeae*, the gonococcus) are typically *intracellular*.

Microscopy. Approximately 1μ in diameter, slightly elongate and arranged in pairs with the long axes parallel and flattening or concavity of their opposed surfaces. Gram-negative, non-motile, non-capsulate and non-sporing; however, when strains of the pathogenic species are freshly isolated capsules may be seen, particularly if specific antiserum preparations are used.

The shape and arrangement of the pathogenic species are usually changed on subcultivation in the laboratory; they become more spherical and tend to appear in clusters.

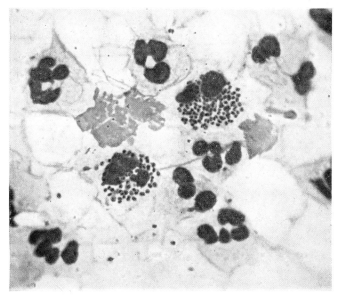

A Gram-stained film of the centrifuged deposit of cerebro-spinal fluid from a case of acute meningitis. Two polymorphonuclear leucocytes are crammed with Gram-negative diplococci; this intracellular appearance is typical of the pathogenic neisseriae and on cultivation a pure growth of oxidase-positive organisms was obtained which on biochemical testing proved to be *N. meningitidis*. ×1000.
This preparation could equally well be that of pus from an acute gonococcal lesion.

Cultural appearances. The pathogenic species are most fastidious and grow best in an atmosphere containing 5 to 10 per cent. CO_2. Blood agar or heated blood agar must be used for primary isolation and incubation at 37°C must be continued for 48 hr or more before one rejects a culture as negative.

Colonies are convex, 1 to 2 mm. in diameter, transparent, non-pigmented and non-haemolytic; considerable variation occurs and prolonged cultivation gives a crenated colony, often with opacity of the central part of the colony and radial striations.

Biochemical activities. ALL neisseriae give a positive **oxidase reaction**; this is tested by applying a freshly prepared 1 per cent. solution of the oxidase reagent (tetramethyl-*p*-phenylenediamine hydrochloride) over the culture plate. Neisseriae turn a rapidly deepening purple colour; it is essential to subculture such colonies within 3 min. of applying the reagent if they are to survive for further study.

Differentiation of the two pathogenic members of the genus from each other and from commensal species is readily performed by simple fermentation reactions; the carbohydrates under test must, however, be prepared in serum broth to satisfy the nutritional demands of the pathogenic members.

Fermentative activities of some Neisseriae

	Glucose	Maltose	Sucrose
N. meningitidis	$\perp$	$\perp$	—
N. gonorrhoeae	$\perp$	—	—
N. catarrhalis	—	—	—
N. sicca	$\perp$	$\perp$	$\perp$

Serological characters. By agglutination tests *N. meningitidis* can be classified into four main groups A to D; group-A strains account for the majority of cases of meningococcal meningitis in epidemics and are regarded as of higher pathogenicity than the other sero-groups which are more usually found in the naso-pharynx of healthy individuals.

N. gonorrhoeae are serologically heterogeneous although *fresh* isolates appear to comprise one group.

COMMENSAL MEMBERS OF THE GENUS. All are oxidase positive and morphologically identical with meningococci and gonococci; all grow readily on ordinary nutrient agar and produce colonies in 18 hr at 37°C. Their colonies are thicker and more opaque than those of the pathogenic species. Some of the commensal members develop highly pigmented colonies.

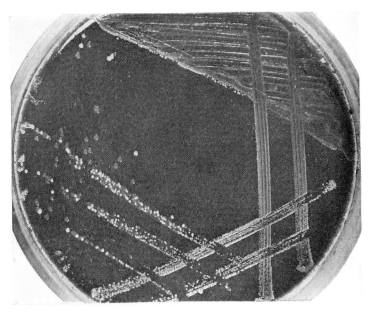

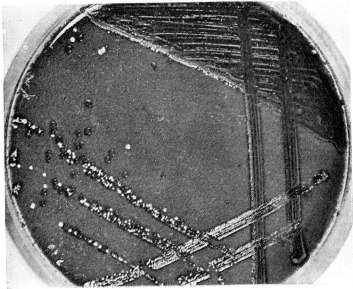

Oxidase reaction

The heated blood agar medium had been inoculated with pus from the eyes
of a two-day-old baby and incubated in an atmosphere of 5 per cent. CO_2
for 48 hr. at 37°C.

The white, pigmented colonies proved to be those of *Staph. epidermidis*
(coagulase −ve); the lower photograph was taken 7 sec. after the oxidase
reagent had been flooded over the plate. The purple-coloured colonies
belong to the genus *Neisseria* and biochemical testing proved them to be
N. gonorrhoeae.

INFECTIONS CAUSED BY NEISSERIAE

As already stated only two members of the genus are pathogenic to man and both are obligate parasites with very poor powers of survival outside the human host.

N. meningitidis is the organism most commonly found in cases of acute pyogenic meningitis and although cases are usually sporadic, epidemic out-breaks are noted from time to time in Britain. In other countries, e.g., Nigeria, wide-spread epidemics are frequent during the dry season.

The meningococcus is rarely involved in other disease conditions but it is occasionally found in cases of conjunctivitis and endocarditis; similarly it can produce a septicaemia without any meningeal involvement and meningococcal septicaemia in an acute or chronic form must be remembered as a possible cause when cases of pyrexia of uncertain origin (PUO) are being investigated.

N. gonorrhoeae reflects its feeble viability outside the host by being almost always associated with adult cases of gonorrhoea which is transmitted by sexual intercourse; in males, acute gonorrhoea usually produces only an acute purulent urethritis but the organisms may spread to the prostate and epididymis. Adult female cases also suffer urethritis and the cervix is also infected and occasionally the infection extends to the endometrium and fallopian tubes.

A bacteraemic phase may occur in both sexes with resultant gonococcal arthritis or rarely endocarditis.

Gonococcal infection of the infant's eyes occurs during childbirth if the mother is suffering from sexually acquired infection but gonococcal ophthalmia neonatorum is very much less common nowadays because of improved ante-natal care and the detection and treatment of gonorrhoea when it occurs in the pregnant woman.

Older children, not yet capable of performing their own toilet and living in hospitals or other institutions may suffer infection of the eyes and, in girls, a vulvo-vaginitis; here the transmission is from an adult attendant suffering from gonorrhoea and who with slovenly habits of personal hygiene passes on the infection by means of damp sponges and towels.

CHAPTER 9

MYCOBACTERIA

THIS genus comprises various types of tubercle bacilli, two of which, the human and bovine types, are important pathogens of man, the leprosy bacillus, and many commensal and saprophytic species.

MYCOBACTERIUM TUBERCULOSIS

Microscopy. Size variable but approximately 3μ by 0·3μ, straight or slightly curved rods with rounded or slightly expanded ends. *Acid and alcohol-fast* (Ziehl-Neelsen stain). Non-motile, non-capsulate, non-sporing.

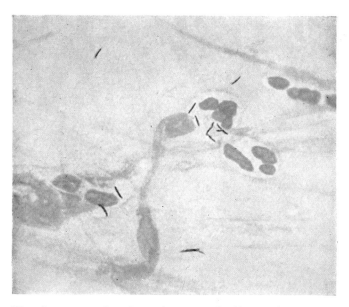

Film of concentrated specimen of sputum stained by the Ziehl-Neelsen method. Acid- and alcohol-fast bacilli can be noted: saprophytic members of the genus are only acid-fast. ×1000.

Cultural appearances. Strictly aerobic. Pathogenic members grow *very slowly* even on rich media such as Lowenstein-Jensen (L-J) egg medium, and macroscopic growth rarely appears until at least 2 to 4 **weeks** incubation at 37°C. The characters of the human type of *M. tuberculosis* on L-J medium can be summarised as 'rough, tough and buff'—i.e., rough=a dry and irregular surface, tough=hard and difficult to emulsify, and buff in colour, compared with the smooth, soft, readily suspended white growth of the bovine type.

The macroscopic appearance of the human type of tubercle bacillus (*M. tuberculosis* var. *hominis*) on Lowenstein-Jensen's medium, grown at 37°C for **6 weeks.** The rough nature of the growth and characteristic colour are apparent.

THREE OTHER TYPES of tubercle bacilli are recognised but only the avian variety has been incriminated (and very rarely) as a human pathogen. This type is morphologically identical with the mammalian types but it has a higher optimum temperature (42°C) and grows more rapidly on L-J medium with production of light brown colonies, highly convex and glistening.

The vole variety is morphologically indistinguishable, non-pathogenic for man and very slow to grow even on L-J medium; particular interest surrounds this variety since it is as efficient in prophylaxis as the more widely used Bacille-Calmette-Guérin (BCG) which is an attenuated bovine strain.

A piscine or cold-blooded type of tubercle bacillus (*M. piscium*) is responsible for a granulomatous condition in fish, frogs, etc., but is non-pathogenic for man. Again these are morphologically similar to other types but have an optimum temperature of 22°C and resemble the avian type in cultural appearances.

Biochemical reactions. These have not been studied systematically.

Serological characters. No valid tests are available for laboratory use; for information on tuberculin preparations a standard text should be consulted.

Animal inoculation. If doubt exists whether an isolate from a patient is of the human or bovine type, the subcutaneous inoculation of a rabbit with the culture allows differentiation; bovine strains kill the animal within a few weeks and post-mortem examination reveals miliary spread. At most, human strains produce only a small lesion at the site of inoculation.

By contrast to the very occasional use of rabbits as outlined above, guinea-pigs are frequently used for diagnostic purposes; intramuscular injection, on the inner aspect of the thigh, with pathological material containing either human- or bovine-type bacilli leads to progressive disease. In communities where disease in the human population is under control, and many cases are 'spotted' early and excrete only few tubercle bacilli, guinea-pig inoculation is more sensitive than *in vitro* cultivation.

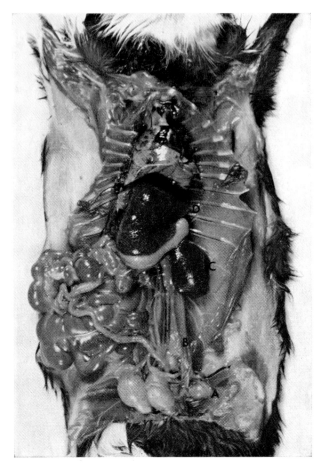

Post-mortem appearance of a guinea-pig sacrificed one month after inoculation with a concentrated specimen of sputum which had not, at that time, shown any growth on Lowenstein-Jensen slopes. The lesions are as follows:

A = Superficial inguinal gland.

B = Para-aortic glands.

C = Slightly enlarged spleen with minute subcapsular tubercles.

D = Liver with subcapsular tubercles.

It is customary to inoculate a pair of guinea-pigs with material from the same individual; this covers the eventuality that one of the pigs may die from intercurrent infection or other disease. The animals are examined daily for evidence of local swelling or of general infection; one of the pair should in any case be killed after 4 weeks and examined for local lesions, involvement of the deep inguinal and para-aortic lymph glands—rarely at this stage is the spleen involved either by enlargement or showing sub-capsular tubercular nodules. If the first animal shows no evidence of disease its partner should be sacrificed after a further 4 to 8 weeks.

When any suspect lesion is noted smears must be made from it and stained by Ziehl-Neelsen's method since guinea-pigs suffer naturally from macroscopically similar lesions due to *Pasteurella pseudotuberculosis* which is *not* acid-fast.

ANONYMOUS MYCOBACTERIA

A great deal of interest has been raised in recent years with the discovery of other members of the genus which are distinct from human- and bovine-type tubercle bacilli but many of which are associated with human disease.

There are four groups of anonymous mycobacteria but members of one group have now been given specific names and are perhaps best spoken of collectively as the intermediate group of mycobacteria since they have lost their anonymity.

Members of this intermediate group are essentially associated with skin infections; *Mycobacterium ulcerans* was originally found in Australia as the cause of indolent skin ulcers and *Mycobacterium balnei*, discovered in Sweden, is associated with spontaneously healing ulcers which originated in areas of skin which had been scratched on the concrete surface of a swimming pool. These organisms can be readily differentiated from tubercle bacilli by their low optimum temperature for growth, 30-33°C, their inability to grow at 37°C and their lack of pathogenicity or feeble pathogenicity for guinea-pigs as well as other characteristics.

These mycobacteria which are truly anonymous can be classified into three groups.

1. *Photochromogens*. Members of this group produce a bright yellow pigment when exposed to light; colonies are smooth and growth is rapid at 37°C (1-2 weeks). Non-pathogenic to guinea-pigs, usually associated with mild pulmonary infections.

2. *Scotochromogens*. Form a yellow pigment even when grown in darkness and when grown in the light an orange pigment is produced; the colonies are smooth and growth is rapid. Organisms are non-pathogenic to laboratory animals. May be found in neck abscesses.

3. *Non-chromogens*. Colonies, which develop rapidly, are colourless although a yellow pigment may be noticeable after exposure to light. Non-pathogenic to guinea-pigs and their association with human disease is usually with an infection similar to pulmonary tuberculosis but its treatment is more difficult since members of the non-chromogenic group of anonymous bacteria are relatively resistant to PAS, INAH and streptomycin.

SAPROPHYTIC AND COMMENSAL MYCOBACTERIA

Saprophytic species are found on grasses and in milk and water; morphologically similar to tubercle bacilli but are not alcohol-fast and grow very rapidly and on ordinary media with production of various pigments. Non-pathogenic to experimental animals.

Particular interest surrounds *Mycobacterium smegmatis* (the smegma bacillus) which is commensal in sebaceous secretions particularly around the prepuce and labia; hence when urine specimens are being collected from suspect cases of genito-urinary tuberculosis careful toilet preparation reduces the number of smegma bacilli in the specimen.

Differentiation of this commensal from pathogenic tubercle bacilli can be made since it is not usually alcohol-fast and grows rapidly on ordinary media; similarly it is non-pathogenic for laboratory animals.

INFECTIONS CAUSED BY TUBERCLE BACILLI

Primary infection with *M. tuberculosis* usually results in a pulmonary lesion called the Ghon focus—this comprises a lesion which is frequently subpleural and also involvement of the hilar lymph nodes. In such instances the human-type bacillus is most commonly involved and infection is by inhalation.

Alternatively bovine-type bacilli ingested in raw milk may settle in the lymphoid tissue of the pharynx and the cervical lymph nodes are also involved or the site of entry may be the lymphoid tissue of the terminal ileum with lymphatic spread to the regional mesenteric lymph glands.

In all such instances the primary infection is frequently subclinical and healing usually occurs by fibrosis and calcification.

Infection with bovine-type bacilli is now uncommon in Britain and in other countries where preventive measures have been taken, i.e., the creation of dairy herds which are free from tuberculosis and/or the pasteurisation of milk has become commonplace.

Post-primary or adult tuberculosis is the infection which is seen clinically and may result either from a freshly acquired exogenous origin, particularly in *young* adults, i.e., under 40 years of age or from a reactivation of an earlier healed lesion as is more likely in older people.

The protective influence of immunisation with BCG is well established as a result of controlled trials carried out in Britain; conclusive evidence is available that BCG or alternatively vole vaccine gives a protection of 80 per cent. against natural infection and that, in particular, these vaccines are especially effective in preventing the more severe forms of human infection, i.e., meningitis and miliary tuberculosis.

Mycobacterium leprae

Probably the causative organism of leprosy; identification is solely on microscopic grounds since the bacillus has not been cultivated nor has the disease been experimentally reproduced in animals.

Microscopy. A Gram-positive bacillus similar in size and shape to the tubercle bacillus; acid-fast when 5 per cent. H_2SO_4 is used for attempted decolorisation. Not alcohol-fast.

Films are prepared from an ulcerated nodule of the skin or material acquired by needle puncture of intact nodules; the presence of large numbers of acid-fast bacilli intracellularly and often in parallel bundles is diagnostic.

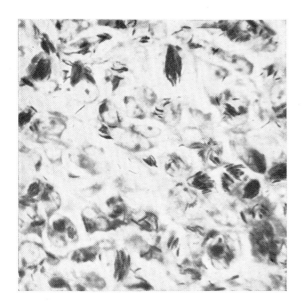

Section of skin showing *M. leprae* as acid-fast rods, mainly in parallel bundles and occurring both intra- and extra-cellularly. ×1000.

F

CHAPTER 10

CORYNEBACTERIA

ONLY one member of this genus, *Corynebacterium diphtheriae*, is pathogenic to man; several species form part of the normal flora in the respiratory tract and other mucous membranes.

C. DIPHTHERIAE

Microscopy. Size very variable, essentially rod-shaped but often showing irregular expansion at one end—'club-shaped'. Cells frequently lie in small clusters at acute angles to each other—Chinese-letter arrangement. Gram-positive but readily decolorised, non-motile, non-capsulate, non-sporing.

Volutin granules can be demonstrated by Albert's staining method in film preparations from *luxuriant* media.

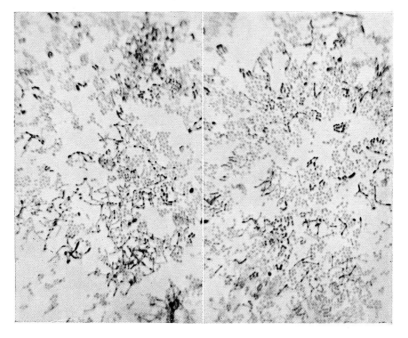

Two films, stained by Albert's method, of material harvested from Loeffler's serum medium. *C. diphtheriae* appear as slender straight or slightly curved bacilli coloured green with volutin granules stained black. The green-stained cocci were noted as *Strept. pyogenes* by parallel cultivation on blood agar. ×1000

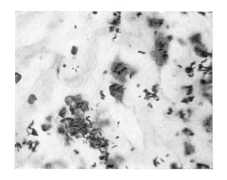

Section of diphtheritic membrane stained by Gram's method showing large numbers of *C. diphtheriae*. ×1000.

Cultural appearances. On Loeffler's serum medium, colonies are small (1 to 2 mm.), circular, grey and with a regular edge after 12 to 18 hr incubation at 37°C; continued incubation gives colonies of larger diameter, with a crenated edge. Volutin granules are abundant in films from Loeffler's medium.

On blood tellurite media the bacilli give grey or black colonies since they reduce the tellurite within the bacterial cell; it is possible to recognise three colony types, *gravis*, *intermedius* and *mitis* on such media.

Volutin granules are very scanty or absent in films made from tellurite media.

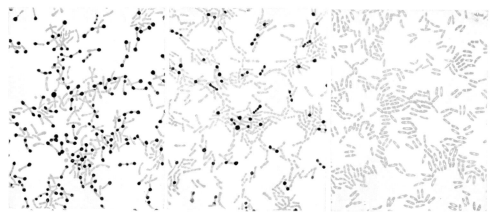

Films, stained by Albert's method, of pure growths of diphtheria bacilli on the left, *C. xerosis* in the centre and *C. hofmannii*. The morphological similarity of *C. diphtheriae* and *C. xerosis* demands that they must be differentiated by further tests; *C. hofmannii* (on the right) shows its characteristic appearance; the absence of volutin granules and the clear unstained central bar are constant features of this organism. ×1000.

Biochemical activities. The activities of the three colonial types of *C. diphtheriae* are contrasted with those of common commensal species in the Table. It should be noted that only *gravis* strains ferment starch.

Biochemical activities of some Corynebacteria

	Glucose	Sucrose	Starch
C. diphtheriae			
var. *gravis*	⊥	—	⊥
var. *intermedius* . . .	⊥	—	—
var. *mitis*	⊥	—	—
C. hofmannii	—	—	—
C. xerosis	⊥	⊥	—

Tube haemolysis tests confirm the identity of the three bio-types of *C. diphtheriae*; *mitis* strains lyse ox and rabbit cells, *intermedius* are inactive against either blood and *gravis* strains lyse only rabbit cells.

Serological characters. The three bio-types are serologically distinct and each can be subdivided into several serological types by agglutination tests. Serotyping studies have been essentially academic but should prove of considerable epidemiological value.

Virulence is equated with the production of a powerful exotoxin and it is essential to demonstrate toxin production in a culture obtained from a throat swab before finally reporting to the clinician. Non-toxigenic strains (=avirulent) are rare in *gravis* (1 per cent.), but more common in *intermedius* (7 per cent.) and *mitis* (15 per cent.) strains.

Animal inoculation. Two biologically equivalent guinea-pigs are used; one of these (control animal) is given an intraperitoneal injection of 1000 units of diphtheria antitoxin 12 hr before inoculation and both guinea-pigs have their abdomens depilated; 0·2 ml. of a 12 hr culture of the organism suspended in broth is then injected *intracutaneously* into each pig.

Virulent bacilli produce a patch of erythema (10 to 15 mm. in diameter) with ensuing oedema and within two days necrosis and then eschar formation follow. The control animal shows no such reaction. If the organism is not toxigenic the test animal shows no reaction; reaction in both test and control animals indicates the organism is not a diphtheria bacillus.

INFECTIONS CAUSED BY DIPHTHERIA BACILLI

Diphtheritic infections of wounds, etc., are occasionally seen but the usual site in which *C. diphtheriae* is found is in the faucial region particularly the tonsils. There is little, if any, lymphatic spread and bacteraemia is unknown; other sites may be parasitised, e.g., the nose, larynx and ear but regardless of site the clinical illness is caused essentially by the diffusible exotoxin which has a predilection for heart muscle and renal tissue. In addition the affinity of diphtheria exotoxin for certain nerves is reflected by post-diphtheritic paralysis of which palatal paralysis is the most common.

Diphtheria was endemic in Britain until 20-25 years ago but was rapidly eliminated by the introduction of mass immunisation with diphtheria toxoid; occasional epidemics still occur but these are almost invariably traced to an imported source.

As with other infections which have been virtually eliminated from Britain by means of active immunisation, the disease is nowadays unknown to the younger generations who not unnaturally are less assiduous in having their children protected by active immunisation; hence a reducing proportion of the population are immune to diphtheria either as a result of natural infection or immunisation. Such a continuing reduction in herd immunity means that the population is increasingly susceptible to epidemic spread should a case occur. Associated with this danger is the fact that the vast majority of younger practitioners have never seen a case of diphtheria and not unnaturally may therefore not consider such a diagnosis so that a missed case or one in which the diagnosis is delayed would not be isolated as quickly and would, therefore, act as a continuing source of infection to other individuals.

CHAPTER 11

BACILLI

(Aerobic, Gram-positive, Spore-forming Rods)

Bacillus anthracis is the only pathogenic member of the genus *Bacillus*. The other members are saprophytic in soil, air, water, etc. *B. anthracis* was the first bacterium proven to be causally related to an infectious disease.

Microscopy. A large organism (4 to 8μ by 1 to $1\cdot5\mu$), rectangular although one or both ends may be concave; tend to occur in chains especially on *in vitro* cultivation. Gram-positive, non-motile, capsulate *in vivo*, spores are central, ovoid and non-projecting and are formed only *in vitro*.

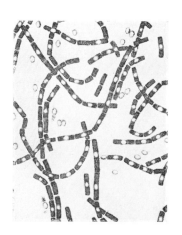

Gram-stained film of culture of *B. anthracis* on nutrient agar; the chains of large square-ended Gram-positive bacilli are typical. Spores which are situated centrally, oval in shape and non-projecting remain unstained whether contained within the vegetative cells or lying free. ×1000.

McFadyean's reaction. Diagnosis of anthrax in animals is usually undertaken by microscopic examination of blood films. The application of polychrome methylene blue for 10 to 20 sec. to a thick blood smear previously fixed by 1:1000 mercuric chloride for 5 min. reveals large blue bacilli lying in a pool of disintegrated heliotrope capsular material.

Cultural appearances. Not in the least fastidious in regard to media, atmosphere or temperature requirements, although aerobic conditions are essential for sporulation and likewise the optimum temperature for spore formation is 25° to 28°C. Colonies on agar or blood agar are circular, have a 'medusa-head' appearance—a colony is one continuous thread of bacilli and has a ground-glass appearance, size 3 mm. in diameter after 18 to 24 hr at 37°C. Only slightly haemolytic on blood agar.

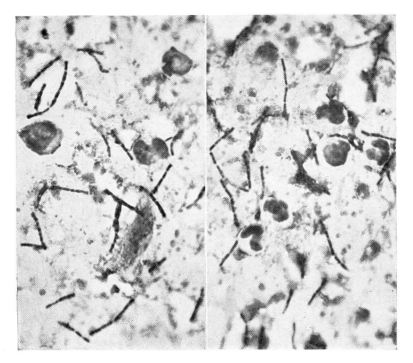

McFadyean's reaction. These blood films, that on the left from the peripheral blood of a cow dying of anthrax and the other from an experimentally infected laboratory animal, have been stained with polychrome methylene blue. Short chains of anthrax bacilli are seen lying among irregular, amorphous, disintegrated capsular material. White blood cells are also present. ×1000.

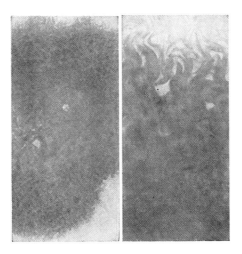

Impression colony of *B. anthracis* stained by methylene blue. The 'medusa-head' appearance is more obvious in the higher power view on the right. ×12, ×75.

Biochemical reactions. Such are rarely employed since the microscopic and cultural appearances of bacilli obtained from an infected animal or human are so characteristic.

Serological characters. The capsules of *B. anthracis* are *not* polysaccharide in composition as are those of many other bacteria, but consist of a polypeptide of D-glutamic acid of high molecular weight which is strongly antigenic; however, anti-capsular antibodies, unlike those of the pneumococci, are without protective effect. Protein and polysaccharide somatic antigens have also been demonstrated but appear to play no greater a role than capsule substance in the natural disease. A lethal toxin has recently been shown as the prime factor in causing death and can be modified for use as an active immunising agent. No typing of *B. anthracis* for epidemiologic purposes has yet evolved.

Animal inoculation. The identity of *B. anthracis* can be confirmed by inoculating a guinea-pig with exudate from the lesion or a culture from such exudate; death with septicaemia occurs within 48 hr and recovery of typical organisms from the heart-blood and spleen is diagnostic. Post-mortem examination of such animals must be undertaken with elaborate precautions to prevent any contamination of the operator or atmosphere and all surfaces must be thoroughly cleaned; disposal of the post-mortem board and carcase is by incineration.

SAPROPHYTIC MEMBERS OF THE GENUS BACILLUS

There are many such and they are usually spoken of as the *Bacillus subtilis* group, the latter organism being the type species. Many of these saprophytes are *motile* and the spores may be central, terminal or sub-terminal; haemolysis on blood agar is common; they do not exhibit McFadyean's reaction and are non-pathogenic for laboratory animals when injected in doses comparable with those of *B. anthracis*.

INFECTIONS CAUSED BY B. ANTHRACIS

Anthrax occurs naturally in nearly all animals and herbivores are the most susceptible; man is infected by spores shed into the environment by sick and dead animals and from spores contaminating animal products, e.g., wool, hides, bone-meal. Infection from another human case is extremely rare. There are obvious occupational risks, e.g., to the farmer and veterinarian.

The internal types of the human disease, i.e., pulmonary and gastro-intestinal anthrax are now extremely rare in Britain, the latter type never having been common. Pulmonary anthrax (wool sorter's disease) has been controlled by treating raw wool before it is processed and also by the provision of local exhaust ventilation which removes wool fibres from the worker's environment.

The external type of human anthrax, i.e., cutaneous anthrax, was once a hazard to porters at docks who humped imported hides but this risk has been virtually eliminated by restricting the importation of hides and other materials and diverting them to ports where they can be mechanically unloaded and processed chemically to remove or destroy spores before they are handled by man.

Again man is protected by strict legislation regarding the safe disposal of any animal which has died from anthrax in Britain.

The commonest source of cutaneous anthrax in this country nowadays is garden fertiliser made from bones contaminated with spores; epidemics of cutaneous anthrax have occurred in various parts of Britain in recent years and have been traced to the handling of bone-meal fertiliser.

Spores enter the skin through abrasions and after 2-3 days a papule appears which enlarges and although there is little or no erythema, the surrounding tissues are disproportionately oedematous compared with the small initial lesion. The painless nature of the lesion coupled with a knowledge of the patient's occupation and/or horticultural zeal should allow differentiation from carbuncles and erysipelas.

Death from cutaneous anthrax has become unusual since penicillin and other antibiotics are effective therapeutic agents.

CHAPTER 12

CLOSTRIDIA

(Anaerobic, Gram-positive, Spore-forming Rods)

The majority of species in the genus *Clostridium* are saprophytic and found in decomposing organic matter, soil, etc.; some are commensal in the intestinal tract of man and animals and a few species are pathogenic. Those causing diseases in man are *Clostridium tetani, Cl. welchii, septicum* and *oedematiens* and *Cl. botulinum*.

Clostridium tetani

This organism, the cause of tetanus in man and many animals, is found in soil, particularly in manured soil and is excreted from the intestinal tract of most animals including man; hence the potential presence of tetanus spores in cat-gut derived from sheep's intestine. In human cases of tetanus, the organisms are localised at the point of entry and the disease is essentially due to the powerful neurotoxin.

Microscopy. Slender rods with round ends, 5μ by 0.5μ, Gram-positive, motile with peritrichous flagella, non-capsulate. Spores are spherical, terminal and project widely—drum-stick bacillus.

Cultural appearances. Strictly anaerobic and rapidly killed by normal atmosphere. Wide temperature range: optimum 37°C. On solid media isolated colonies are rarely seen since the organism spreads as a diaphanous film which might escape brief examination; the edge of the spreading growth shows numerous projections.

Grows readily in cooked-meat medium with only slight digestion.

Tube of nutrient agar inoculated throughout its depth with a straight wire loaded with *Cl. tetani*, and incubated for 48 hr at 37°C. The strictly anaerobic nature of the organism is shown by the absence of surface growth and the increasing width of the lateral spikes in the deeper parts of the medium—'fir-tree' growth.

Blood agar plate viewed by oblique illumination. *Cl. tetani* had been inoculated in a diametric streak and after 24 hr incubation in an anaerobic jar (37°C) the original inoculum line shows confluent growth; above this and bordering on the reflection area is the very fine, diaphanous film of spreading growth which is difficult to detect with the naked eye.

Biochemical activities. No sugars are fermented; indole is produced.

Serological characters. Ten serotypes are distinguished on the basis of flagellar antigens. In all types, an identical neurotoxin is produced, but some otherwise typical strains are non-toxigenic.

Animal inoculation. This is the most reliable laboratory technique for diagnosis. A pair of mice is used for each test; one mouse is protected by subcutaneous injection of 750 units of tetanus antitoxin 2 hr before inoculation. Both animals are inoculated intramuscularly in the right hind leg; the inoculum comprises 0·25 ml. of supernatant from a 48-hr cooked-meat culture of the organism. Within a few hours, evidence of tetanus develops in the unprotected mouse—the tail becomes stiff, the inoculated limb progressively paralysed; thereafter the paralysis becomes more generalised and tetanic convulsions can be elicited by slight stimuli.

CLOSTRIDIUM WELCHII (CL. PERFRINGENS)

This organism is the commonest of several members of the genus *Clostridium* associated with gas gangrene. The main source of *Cl. welchii* is the excreta of animals, including man. It must be appreciated that, as in the case of tetanus, the diagnosis of gas gangrene is made on clinical grounds since the presence of *Cl. welchii* in a wound is, in itself, of no significance; classically gas gangrene is a rapidly spreading infection with oedema, necrosis, gas production and tissue gangrene.

Microscopy. Relatively large bacilli, 5μ by 1μ, with square or rounded ends. Gram-positive, non-motile (all other *Clostridia* are motile), capsulate in animal tissues. Spores are oval, subterminal and non-projecting.

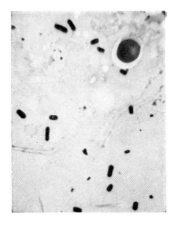

Two Gram-stained preparations showing *Cl. welchii*. On the left is a smear made directly from muscle in a case of gas gangrene; on the right is a section of muscle from the same case. Separation of the muscle fibres, which are also oedematous, by gas production can be be noted. ×1000, ×500.

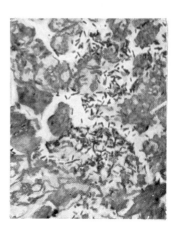

Cultural appearances. Not as strictly anaerobic as *Cl. tetani*; grows very rapidly. Colonies on blood agar are approximately 3 mm. in diameter after 18 to 24 hr incubation, semi-translucent and with an entire edge. Many strains show zones of β-haemolysis.

Biochemical activities. Ferments glucose, lactose, maltose and sucrose with much gas production. Indole is not produced. In litmus milk, rapid fermentation of the lactose occurs with the appearance of a 'stormy clot' reaction.

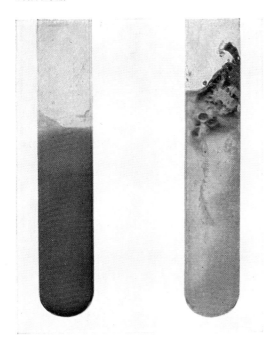

Stormy clot reaction. On the left is an uninoculated tube of litmus milk; that on the right has been inoculated with *Cl. welchii* and incubated anaerobically for 24 hr at 37°C. The milk has been clotted with acid formation, and gas production has broken up the clot. The yellowish fluid is the whey of the milk. The stormy clot reaction is not as highly specific for *Cl. welchii* as is sometimes believed since certain other members of the genus give similar reactions in litmus milk.

Serological characters. Five types (A-E) may be differentiated according to the exotoxins produced; type A is that most commonly associated with gas gangrene, the other types are more commonly associated with animal diseases.

NAGLER'S REACTION. All types of *Cl. welchii* produce opalescence in egg-yolk media, due to production of lecithinase which causes a visible precipitate. Such opalescence is specifically inhibited by *Cl. welchii* antitoxin; thus plates of the medium with one half of the surface coated with antitoxin and then inoculated on both halves allows identification of the organism. Several other *Clostridia* give similar reactions on Nagler plates and inhibition of opalescence can be demonstrated with their specific antisera.

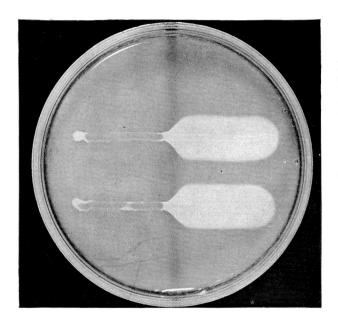

Nagler's reaction. The medium comprises nutrient agar with 5 per cent. egg yolk. Three drops of *Cl. welchii* type-A antitoxin were smeared over the left half of the plate as marked by the blue line; a culture of *Cl. welchii* was streaked in two parallel lines across the surface of the medium and the plate then incubated anaerobically at 37°C for 20 hr. Lecithinase activity produces zones of opacity on the antitoxin-free half of the plate and this activity has been inhibited in the presence of antitoxin on the left half.

Of the *Clostridia* considered in this chapter *Cl. botulinum* (all types) and *Cl. oedematiens* (except type C) produce opalescence but in the case of these other organisms the opalescence is not inhibited by specific antiserum except for that produced by *Cl. oedematiens* type A.

Animal inoculation. Virulence varies with different strains; with pathogenic strains the intramuscular injection of 1 ml. of a fresh 24-hr cooked-meat broth culture into a guinea-pig causes death in 1 to 2 days. Within a few hours of injecting a limb, marked oedema occurs and frequently crepitation can be detected; oedema and gas formation spread rapidly and the organisms are recoverable from the heart-blood within 12 hr of inoculation. The administration of *Cl. welchii* antitoxin before injecting the culture fluid protects a control guinea-pig.

Of the several other Clostridia associated with gas gangrene, *Cl. oedematiens* and *Cl. septicum* occur most frequently.

Robertson's cooked-meat medium. The centre tube was not inoculated while that on the left was inoculated with *Cl. welchii*; the meat is not digested but is slightly reddened due to saccharolytic activity. A similar appearance results with *Cl. oedematiens* and *Cl. septicum*. The right-hand tube was inoculated with an essentially saprophytic species, *Cl. sporogenes*, which has effected gross digestion of the meat particles by proteolytic activity.

CLOSTRIDIUM OEDEMATIENS (CL. NOVYI)

Microscopy. In stained films two features serve to differentiate *Cl. oedematiens* from *Cl. welchii;* these are the projecting nature of the spores of the former and the fact that they usually stain Gram negatively. It is, however, motile and non-capsulate.

Cultural appearances. Strictly anaerobic and all strains tend to show spreading growth on solid media; of the four sero-types, A to D, all except those of type C are haemolytic on blood agar. Serotype B and D strains require the presence of cysteine and dithiothreitol for reliable surface growth to take place.

Biochemical activities. Fermentative reactions are wide and may be variable although glucose and maltose are utilised by all types except type D which ferments only glucose and contrarily this is the only type which produces indole.

Serological characters. Types A, B and D produce exotoxins which are antigenic although cross-reactions with specific antitoxins can occur. Type C does not produce toxin in culture; type A strains can cause gas gangrene in man whereas type B and D strains predominantly affect animals.

Animal inoculation. Guinea-pig inoculation and protection tests can be performed in a manner similar to those for *Cl. welchii* but cross reactions between the serological types may make interpretation difficult.

CLOSTRIDIUM SEPTICUM

Microscopy. Tend to be more slender than *Cl. welchii* and citron-shaped organisms or filamentous forms occur. Spores are oval, subterminal and projecting.

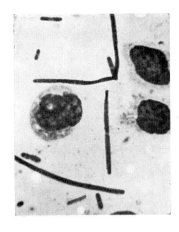

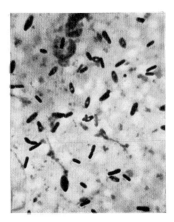

Gram-stained preparations of *Cl. septicum.* On the left, bacillary and filamentous forms can be seen, whereas on the right, citron-shaped forms predominate. In the latter preparation a few organisms show the presence of oval, projecting spores situated subterminally. ×1000.

Cultural appearances. Strictly anaerobic. Colonies are 3 mm. in diameter after 48 hr incubation, irregularly circular with a filamentous border. Growth tends to spread over the medium.

Biochemical activities. Glucose, maltose, lactose and salicin are fermented. Indole is not produced.

Serological characters. Four groups are recognised on the basis of somatic antigens and each group is further divided by the possession of various flagellar antigens.

Animal inoculation. Guinea-pigs are susceptible to inoculation and can be protected with antitoxin.

An almost identical organism, *Clostridium chauvoei*, is pathogenic only for animals. Biochemically it ferments sucrose and not salicin; serological relationships with *Cl. septicum* are recognised.

CLOSTRIDIUM BOTULINUM

This organism, essentially a saprophyte, is widely distributed in nature and quite frequently present in the intestinal tract of domestic animals. Botulism in man is an intoxication resulting from ingestion of foodstuffs containing toxin produced by the organism growing in the food.

Microscopy. Large, straight-sided rods some 4μ by 1μ in size. Gram-positive, motile, non-capsulate. Spores are oval, central or subterminal in position and projecting—not unique in any morphological respect.

Gram-stained film from culture of *Cl. botulinum.* A few organisms show clear unstained areas occupied by spores which are oval, subterminal and projecting. ×1000.

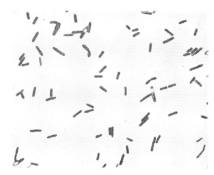

Cultural appearances. Strictly anaerobic—there is much variation in colonial morphology but colonies are often irregularly round with a fimbriated edge, some 3 mm. in diameter, translucent. Optimum temperature 35°C.

Biochemical activities. The six types—A to F—vary in their saccharolytic and proteolytic activities.

Serological characters. The six types produce neurotoxins which are antigenically distinct so that specific antitoxins can be used in guinea-pig tests to determine the type of the isolate.

Animal inoculation. Botulism can be demonstrated by intra-peritoneal injection of a fluid culture into a guinea-pig; a few hours after inoculation dyspnoea occurs followed by flaccid paralysis of the abdomen and ultimately generalised paralysis so that the animal becomes quite inert. As with similar tests for *Cl. tetani* paired animals are used, one protected with polyvalent botulinus antitoxin. Determination of the type of neurotoxin can be made by employing additional guinea-pigs each protected with type-specific antitoxin to the five types of neurotoxin.

Man is only affected by *Cl. botulinum* types A, B and E.

Biochemical reactions of certain Clostridia

	Glucose	Lactose	Maltose	Sucrose	Salicin	Indole
Cl. tetani	—	—	—	—	—	+
Cl. welchii	+	+	+	+	+	—
Cl. oedematiens	+	—	+(D)	—	—	—(D)
Cl. septicum	+	+	+	—	+	—
Cl. botulinum	+	—	+	—	+{ (B) (C)	—

D=*Cl. oedematiens* type D does not ferment maltose but does produce indole.
B and C=*Cl. botulinum* types B and C do not ferment salicin.

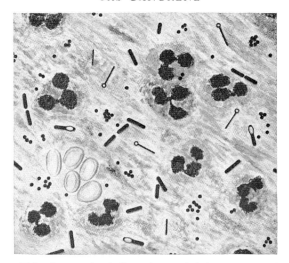

This Gram-stained film of material from a wound shows pus cells (degenerate polymorphonuclear leucocytes), a few necrotic muscle fibres and clumps of Gram-positive cocci which on culture proved to be coagulase positive staphylococci.

Two species of *Clostridia* were also isolated, *Cl. tetani* which in the film is represented by the slender Gram-positive rods and also rods with spherical, terminal and projecting spores which have stained Gram-negatively; such variation in reaction to Gram's staining method is well recognised. The stouter Gram-positive bacilli, some of which bear oval, sub-terminal, projecting spores were finally identified as *Cl. oedematiens*.

INFECTIONS CAUSED BY CLOSTRIDIA

Gas gangrene. Although *Cl. welchii* is the most frequently encountered species in gas gangrene other members of the genus, and particularly *Cl. oedematiens* and *Cl. septicum*, are also incriminated in this infection; more than one species is frequently found in cases of gas gangrene.

Infection results from the contamination of a wound by soil, dirty clothing, etc., but not all wounds from which such organisms are recovered develop gas gangrene and the diagnosis and initiation of treatment therefore must remain the responsibility of the clinician; however, endogenous infection is common in civilian practice because faecal carriage of *Cl. welchii* is the rule.

On clinical grounds we can recognise three types of involvement of gas gangrene organisms: firstly a simple contamination of a wound as noted above with healing taking place without any evidence of infection; alternatively the organisms may multiply and spread along intermuscular septa but without invading uninjured, healthy muscle. The term anaerobic cellulitis is used for this localised condition.

Finally the organisms may spread readily and with extremely invasive powers into healthy muscle which is destroyed by the powerful exotoxins; this is classical gas gangrene or clostridial myonecrosis.

Active immunisation against gas gangrene is not practicable because of the various species involved; polyvalent antiserum is available for prophylactic emergency use and for therapy. In an established case and if the causal organism has been identified several monovalent sera are available for therapeutic use but neither serotherapy nor antibiotic therapy can replace adequate surgical treatment.

Cl. welchii food-poisoning. Certain strains of *Cl. welchii* can cause toxic food-poisoning; such strains occur in the faeces of man and animals and are widely distributed. The vehicle of infection is almost invariably some pre-cooked meat which has been stored at temperatures which allow multiplication of *Cl. welchii*; following on ingestion of such food, containing a sufficiently large number of organisms, the patient suffers abdominal discomfort and cramps some 8-14 hr later. This is followed by diarrhoea and the illness, which is brief and lasts only 1-3 days, is usually apyrexial.

Tetanus. This infection follows on the implantation of tetanus spores in a wound where conditions are suitable for their germination and growth of the vegetative cells.

As already stated tetanus bacilli are commensal in the intestinal tract of many mammals including man and tetanus spores are thus ubiquitous and, in particular, they are found in large numbers in manured soil.

The clinical condition is caused by the powerful exotoxin, *tetanospasmin*, produced by vegetative cells in the injured area; the bacilli remain localised in the wound.

Tetanus neonatorum follows infection of the umbilical stump and this form of tetanus is most common in communities where cow dung and such like are used as umbilical dressings; it can, however, occur in more sophisticated communities if, as a result of inadequate sterilisation, tetanus spores survive in cord dressings or talc used to powder the umbilical area. Although tetanus occurs classically in the deep, dirty wound contaminated with soil and with necrotic tissues many cases occur in some countries, e.g., Nigeria, where there is no obvious injury; it is considered that in these instances the spore is implanted in minor cracks or abrasions on the soles of the feet in those who walk bare-footed.

Similarly in Britain cases occur in which the trauma is minimal, e.g., a thorn prick or a simple superficial skin abrasion; such instances underline the need for the much wider use of active immunisation with tetanus toxoid; such a practice would also reduce the use of tetanus antitoxin for emergency passive immunisation in the injured and thus lower the risks of hypersensitivity reactions which are a recognised hazard of the administration of this and other antiserum preparations.

In any case, and even in individuals who have been actively immunised, adequate surgical treatment of all wounds remains the sheet anchor of tetanus prophylaxis.

Botulism. Fortunately this disease is rare in Britain; it results from the ingestion of foodstuffs, particularly preserved foods, in which *Cl. botulinum* has been growing and produced its highly lethal exotoxin which affects specifically the parasympathetic nervous system. Although the exotoxin is most potent its clinical effects are not evident until 12-36 hr after the incriminated food has been ingested and occasionally even 3 or 4 days may elapse before symptoms appear; in general, the shorter the incubation period the more severe the disease is and the higher the fatality; death is due to cardiac or respiratory failure.

Prevention of botulism demands that preserved foods must be adequately prepared and stored and home-canning in particular should be avoided; toxoid preparations are available for active immunisation but the very low incidence of the disease does not justify their use in man.

CHAPTER 13

THE ENTEROBACTERIA

THE family *Enterobacteriaceae* comprises numerous inter-related genera all of which are microscopically indistinguishable, 3 to 5µ by 0.5µ, relatively straight rods with rounded ends, Gram-negative, variously motile, some are capsulate, none possesses spores. All members of the family ferment glucose, with or without gas production. Their natural habitat is the intestinal tract of man and animals; some, e.g., *Escherichia coli*, are part of the normal flora whilst others, e.g., shigellae and salmonellae, are pathogenic for man.

Not all of the genera of this family are considered in this volume.

ESCHERICHIA COLI

Microscopy. Strains have the general characteristics mentioned above; those which are motile possess peritrichous flagella.

 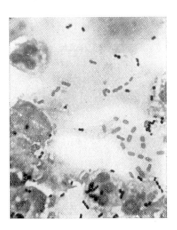

On the left is a Gram-stained film of a culture of *Esch. coli;* the organisms are morphologically indistinguishable from other enterobacteria, whether these are commensal or pathogenic. On the right is a Gram-stained film from a centrifuged deposit of urine and the appearance of *Esch. coli* can be contrasted with that of *Strept. faecalis.* × 1000.

Cultural appearances. Grows readily on all ordinary media and is aerobic and facultatively anaerobic. Colonies are 2 to 4 mm. in diameter after 18 to 24 hr at 37°C, opaque and convex with an entire edge.

On MacConkey's medium colonies are rose-pink on account of lactose fermentation; grow poorly if at all on desoxycholate citrate agar (DCA) with small dense pink colonies.

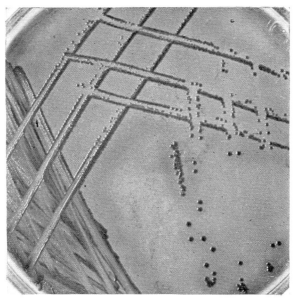

MacConkey's plate after overnight incubation at 37°C. The plate had been seeded with pus from a pilonidal sinus; characteristic lactose fermenting colonies of *Esch. coli* in profuse, pure culture.

Biochemical activities. The practical importance of these tests lies in differentiating *Esch. coli* strains, which are constantly present in human faeces, from the microscopically similar common pathogenic species of *Shigellae* and *Salmonellae*. More than 95 per cent. of *Esch. coli* strains ferment lactose promptly and with gas production so that confusion with shigella strains rarely occurs. Virtually all strains produce indole so that this feature alone prevents wrong identification as a member of the genus *Salmonella* which never produces indole; the absence of urease activity separates *Esch. coli* from members of the genus *Proteus* at an elementary level. As in the identification of any species within the *Enterobacteriaceae* the final decision rests on serological differentiation.

Serological characters. This genus has been subjected to extensive antigenic analysis and more than 140 somatic (O) antigens have been identified and 49 flagellar (H) antigens are known. Routine serotyping is not undertaken except when strains are implicated in gastroenteritis; these enteropathogenic strains possess K antigens—present in capsules or microcapsules. Three types of K antigen, L, A and B, can be differentiated by their stability in various physical tests. Almost all enteropathogenic strains possess the B type of antigen.

Animal inoculation is not undertaken in diagnostic bacteriology but many strains are associated with natural disease in domestic animals and poultry.

BIOCHEMICAL IDENTIFICATION OF ENTEROBACTERIA

For many years the biochemical differentiation of the common bowel pathogens, shigellae and salmonellae, from culturally similar non-pathogenic species has been made by inoculating colonies into a differential row of sugars (p. 37). By using two composite media, preliminary identification of shigella and salmonella organisms can be effected with economies in time, labour and glassware.

The pair of composite media shown at A are uninoculated; tube 1 contains a mixture of glucose, mannitol and urea in a normal concentration of agar and is prepared as a slope with a deep butt. Tube 2 contains a mixture of sucrose and salicin in semi-solid agar over which lead acetate and Kovac reagent impregnated papers are suspended. Both media also contain relevant indicators.

The media are inoculated with a long, straight wire charged with a pure culture of the organism to be identified. Tube 1 is inoculated by both smearing the slant and then stabbing to the base of the butt; tube 2 is then inoculated by a single stab into its upper $\frac{1}{2}$ in. after which the test papers are suspended above the medium and held by the cotton-wool stopper.

The tubes shown at B reveal that the organism with which they were inoculated produces urease by the first tube showing a deep blue colour and also produces indole (yellow test paper) and utilises sucrose and/or salicin—such a pattern indicates that the culture was of the genus *Proteus*.

The reactions produced by *Shigella sonnei* and indole non-producing types of *Sh. flexneri* and *Sh. boydii* are shown in tubes C; glucose and mannitol have both been utilised in the first tube and without gas production; these species do not attack sucrose or salicin and are non-motile so that their growth in the second tube is restricted to the original inoculum track.

The first tube at D shows gas production accompanying glucose and mannitol fermentation and in tube 2 motility is indicated by diffusion of the strain from the original inoculum line throughout the semi-solid medium; H_2S has been produced and is indicated by blackening of the lead-acetate paper. Indole has not been produced and no fermentation has occurred; the pattern of the reactions is biochemically consistent with a member of the genus *Salmonella*, other than *S. typhi* which is anaerogenic.

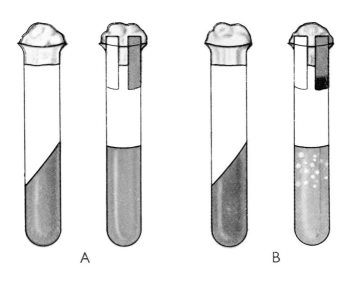

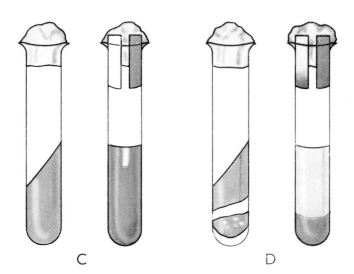

Composite media for preliminary biochemical differentiation of
Shigella and Salmonella.

SHIGELLAE

Members of this genus are the causative organisms of acute bacillary dysentery which is world-wide in distribution and endemic even in many highly developed countries including Britain. With the exception of certain simians man is the only host.

Microscopy. Identical with *Esch. coli* but members of the genus *Shigella* are *never* motile or capsulate.

Cultural appearances. Similar to *Esch. coli*, but give pale (colourless) colonies on MacConkey's and DCA medium since, with the exception of *Shigella sonnei*, they do not ferment lactose. Shigellae grow abundantly on DCA in comparison with *Esch. coli*.

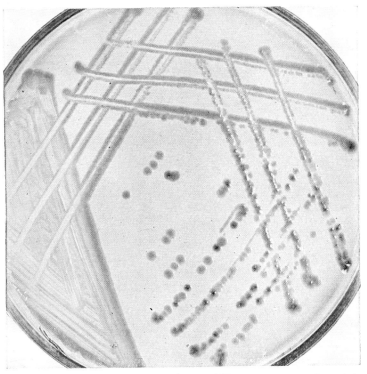

MacConkey's plate sown with a mixture of *Esch. coli* (scanty, deep pink colonies) and *Sh. sonnei*; the plate has been incubated for 24 hr at 37°C. For this reason the colonies of *Sh. sonnei* show a slight colour change from the pale, colourless state characteristic of shorter periods of incubation. *Sh. sonnei*, the commonest cause of bacillary dysentery in Britain, is a late lactose fermenter; all other members of the genus are lactose non-fermenting.

Biochemical activities. Four groups are readily differentiated as in the Table.

Groups within the genus Shigella

	Lactose	Mannitol	Indole	No. of Serotypes
A. *Sh. dysenteriae*. .	—	—	V	10
B. *Sh. flexneri* . .	—	⊥	V	6
C. *Sh. boydii* . .	—	⊥	V	15
D. *Sh. sonnei* . .	(⊥)	⊥	—	1

Group A strains are mannitol non-fermenting; members of groups B and C must be differentiated by serological methods but it should be noted that *Sh. flexneri* serotype 6 strains are capable of biochemical subdivision and that certain of these are exceptional in producing gas (in small volumes). Group D strains are unique in fermenting lactose if incubated beyond 18 to 24 hr and thus give pink colonies on MacConkey and DCA media.

Serological characters. Each of the groups is serologically distinctive; 10 serotypes are recognised within Group A and there is no significant intra-group relationship. Group B comprises six serotypes, all of which possess a common group antigen and each possesses a type-specific component.

Sh. flexneri types 1, 2 and 4 can be divided each into two subtypes on the presence or absence of minor group antigens. Group C contains 15 serotypes which are serologically distinct from Group B strains in lacking the *flexneri* group antigen. There are intra-group relationships so that for identification of the serotype of a *Sh. boydii* strain it is necessary to use absorbed antisera. Group D strains are a single serological entity. *Shigella dysenteriae* serotype 1 (*Sh. shigae*) produces a potent neurotoxin.

Animal inoculation. Of no value in the diagnostic identification of shigellae.

COLICINE TYPING

Colicines are naturally occurring antibiotics produced by many members of the family Entero-bacteriaceae; the colicine activity of a strain may be directed against other genera within the family although usually such activity is most commonly displayed against members of the same genus as that of the producing strain.

Use can be made of colicine production to characterise some organisms by noting various patterns of inhibition on a collection of passive or indicator strains. This method is proving helpful in elucidating the spread of *Shigella sonnei*; this is the organism most commonly incriminated in cases of bacillary dysentery in Britain, an infection which has shown an almost unrelenting increase in incidence over the last thirty or more years.

Some of our inability to control this disease may have been associated with the fact that there is only one serological type of Sonne's bacillus so that all strains isolated from cases are serologically identical. We can now, by means of colicine typing, differentiate at least seventeen stable types so that any particular type can be traced as it spreads through the population.

TYPING TECHNIQUE

Production. The strain to be typed, i.e., the producer strain, is diametrically streaked on tryptone soya blood agar. The plate is then incubated for 24 hr at 35° C; the temperature of incubation is important and should not be allowed to rise or fall beyond 1° C from the stated optimum, otherwise certain strains give aberrant results.

Processing. After this period of preliminary incubation the macroscopic growth of the producer strain is removed with a microscope slide and the microscopic remnants of growth are sterilised by exposing the surface of the plate to chloroform vapour for 10 min. Thereafter the plate is exposed to air for a few minutes to get rid of all traces of chloroform.

The 15 passive (indicator) strains are streaked on to the plate at right angles to the original growth line of the producer strain. Eight strains are inoculated on the left hand and the balance of seven strains on the right-hand side of the plate; this distribution serves as a marker for the indicator strains.

Interpretation. A further period of incubation at 37° C (the temperature is not now important) will allow any colicines which have diffused into the medium during growth of the producer strain to exert their antibiotic activity on particular indicator strains. The antibiotic activity of *Sh. sonnei* type 7 is shown diagrammatically and it will be noted that all of the indicator strains have grown with the exception of strain number 3 which has been fairly widely inhibited. The accompanying photographs show the patterns of inhibition given by a type 7 producer strain, a type 3 producer strain and also the activity of a producer strain of *Sh. sonnei*, type 2.

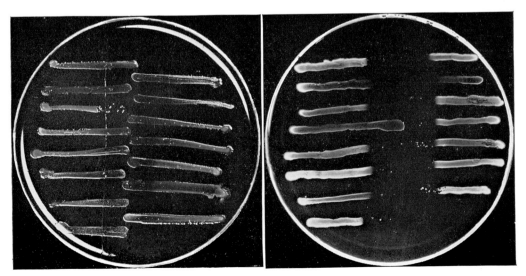

Type 7   Type 3

OUTLINE OF COLICINE-TYPING TECHNIQUE FOR SH. SONNEI.

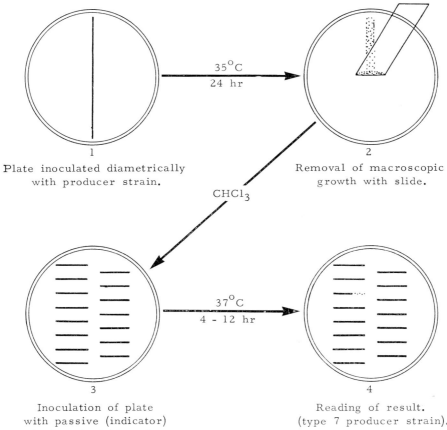

1
Plate inoculated diametrically
with producer strain.

35°C
24 hr

2
Removal of macroscopic
growth with slide.

CHCl$_3$

3
Inoculation of plate
with passive (indicator)
strains at right angles
to original growth line.

37°C
4 - 12 hr

4
Reading of result.
(type 7 producer strain).

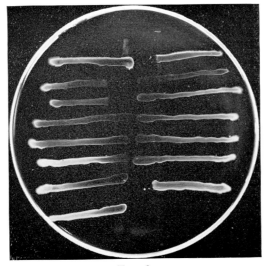

Type 2

95

SALMONELLAE

There are more than 1000 serotypes within the genus *Salmonella*; four of these, *Salmonella typhi*, *S. paratyphi A*, *S. paratyphi B* and *S. paratyphi C* give rise to the enteric fevers and are parasitic only in the human intestine. The other serotypes occur widely as parasites of mammals and birds as well as of man and the human infection associated with them is in the nature of acute gastroenteritis or bacterial food-poisoning.

Microscopy. Identical with *Esch. coli* but with the exception of *S. pullorum* and *S. gallinarum*, all types are motile and possess peritrichous flagella.

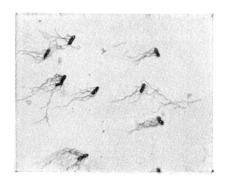

Film of culture of *S. typhi* stained by a silver deposition method which thickens the flagella at least tenfold thus allowing their demonstration by the light microscope. ×1000.

Cultural appearances. Aerobic and facultatively anaerobic; have a wide temperature range and like all enterobacteria grow readily on all ordinary media. Colonies are 2 to 4 mm. in diameter after 18 to 24 hr incubation at 37°C and on agar appear as greyish-white dome-shaped discs with an entire edge. On MacConkey's and DCA media colonies are similar in size and shape to those on agar and are pale since lactose is not fermented.

DCA plate which had been inoculated with faeces from a case of bacterial food poisoning and then incubated overnight at 37°C. Pale, lactose non-fermenting colonies can be seen lying among the pink colonies of lactose fermenting species; the pale colonies were provisionally identified as belonging to the genus *Salmonella* by biochemical and motility tests. Subsequently, serological examination revealed that they were *S. typhimurium*. ×2. (This culture plate was photographed against a dark background, thus the colour of the medium is not apparent.)

Biochemical activities. There is a profusion of biochemical tests but for normal diagnostic purposes it is sufficient to determine that the organism ferments glucose, dulcitol and mannitol with gas production (N.B.—*S. typhi* is anaerogenic), does not ferment lactose or sucrose and does not produce indole. On this evidence and the demonstration of motility further identification is by serological analysis.

Serological characters. All motile salmonellae possess two main antigens. The 'O' or somatic antigens are group antigens and it will be noted in the Table that identification of these with group-specific antisera allows the unknown organism to be allocated to one or other of the groups. The 'H' or flagellar antigens of a single species may occur in either or both of two phases; phase 1 antigen being shared by only a few other species whereas phase 2 is shared by many; these phases are referred to respectively as specific and non-specific phases. A few salmonellae possess a third antigen occurring on the surface and designated the Vi antigen; when present it masks agglutination with 'O' antisera.

Salmonellae. Some representatives of the genus

(Kauffman-White schema)

Group	Serotype	Somatic antigens	Flagellar antigens	
			Phase 1	Phase 2
A.	*S. paratyphi A*	1, 2, 12	a	—
	S. kiel	1, 2, 12	g, p	—
B.	*S. paratyphi B*	1, 4, 5, 12	b	1, 2
	S. typhimurium	1, 4, 5, 12	i	1, 2
C_1.	*S. paratyphi C*	6, 7, Vi	c	1, 5
	S. thompson	6, 7	k	1, 5
C_2.	*S. newport*	6, 8	e, h	1, 2
	S. bovis-morbificans	6, 8	r	1, 5
D.	*S. typhi*	9, 12, Vi	d	—
	S. enteritidis	1, 9, 12	g, m	—

BACTERIOPHAGE TYPING. Certain salmonella serotypes can be subdivided into phage-types for epidemiologic purposes.

Animal inoculation. Species vary in their virulence for various laboratory animals; inoculation is of no importance in diagnostic laboratory practice.

PROTEUS

Members of this genus are widely distributed in nature and are also found in the faeces of animals and man. They occur as pathogens in wounds, bed-sores and urinary tract infections; infection may be exogenous or endogenous in origin.

Microscopy. Similar to other enterobacteria but very pleomorphic. Motile.

Cultural appearances. Not in the least fastidious in regard to temperature, atmosphere or nutritional requirements. On nutrient agar these organisms rarely grow as isolated colonies but swarm in successive waves over the surface; swarming is inhibited on MacConkey's medium or by increasing the agar content of blood agar to 4 per cent.

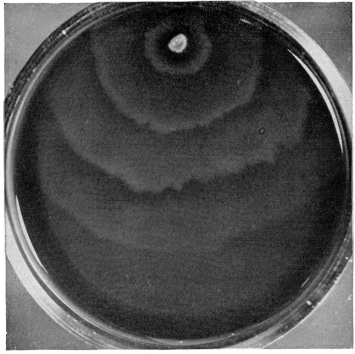

Blood agar plate which had been inoculated at one point with *Proteus vulgaris*. After 18 hr incubation at 37°C four successive waves of growth can be seen; swarming of Proteus species delays the isolation of other organisms in mixed cultures but this characteristic can be inhibited by incorporating various substances in the medium, e.g., sodium azide (1 in 5000), chloral hydrate (1 in 500) or by increasing the agar content. Similarly the inhibition of swarming on MacConkey's medium is valuable in allowing separation of other species provided that these grow on this selective medium.

Biochemical activities. Four biochemical types can be recognised as shown below.

Biochemical types of Proteus

	Mannitol	Maltose	Indole
Pr. vulgaris . . .	—	+	+
Pr. mirabilis . . .	—	—	—
Pr. morganii . . .	—	—	+
Pr. rettgeri	⊥ or +	—	+

Lactose is not fermented so that on MacConkey or DCA media, pale colonies are noted. Differentiation from the popular pale pathogens isolated on such media is effected by testing for urease activity. Proteus species decompose urea rapidly with the liberation of ammonia; *Shigellae* and *Salmonellae* do not produce urease.

Serological characters. *Pr. vulgaris* and *Pr. mirabilis* have been subjected to serological analysis and 119 serotypes can be recognised on the basis of O and H antigens, analogous to serotyping of salmonellae.

Certain types of proteus have antigens in common with the rickettsiae and this is the basis of the Weil-Felix reaction in the sero-diagnosis of the typhus fevers.

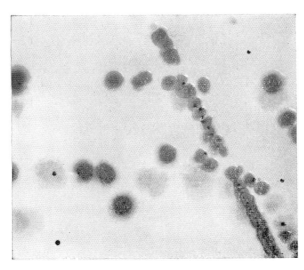

This plate of MacConkey's medium was inoculated with a specimen of urine from a with patient paraplegia of several years' duration; the large and small lactose fermenting colonies are those respectively of *Esch. coli* and *Strept. faecalis*. The pale, lactose non-fermenting colonies are those of *Pr. mirabilis;* note the inhibition of swarming of *Pr. mirabilis* on this medium. ×3.

KLEBSIELLAE

The majority of types within this genus are saprophytic, e.g., in water supplies. Other strains occur commensally in the intestinal tract in a minority of healthy people. The most common site in which klebsiella strains fulfil a pathogenic role in man is the urinary tract.

Microscopy. Similar to other members of the *Enterobacteriaceae* but are non-motile. Capsulate both in the tissues and on *in vitro* cultivation.

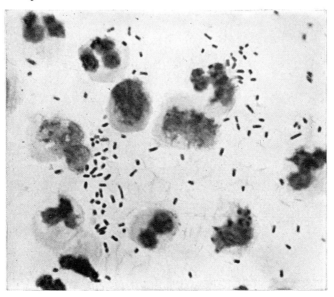

Gram-stained film of urinary deposit showing Gram-negative bacilli and polymorphonuclear leucocytes; some of the latter are disintegrating. The bacilli, which are morphologically indistinguishable from other enterobacteria by Gram's method, were shown to belong to the genus *Klebsiella*. ×1000.

Cultural appearances. Colonies are large, high-convex and mucoid on account of abundant production of extracellular slime; colonies tend to coalesce. On MacConkey's medium the majority of strains give pink colonies due to lactose fermentation.

Biochemical activities. Strain variation is so great that it defies summary treatment—in any case their characteristic colonies allow easy differentiation from other enterobacteria. The majority of strains are urease-producers but are much slower and less intense in this regard than proteus strains; their lack of motility and non-spreading growth on ordinary media further differentiate them from proteus.

Serological characters. In addition to somatic antigens, strains also possess K (capsular) and M (mucoid) antigens; in any one strain the K and M antigens are identical. The presence of K and M antigens masks the O antigen so that capsular antisera are used to characterise strains. Thus, the genus is divided into 72 defined serotypes by means of capsule-swelling reactions similar to the technique employed in typing pneumococci.

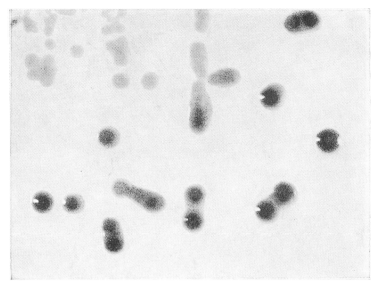

This MacConkey plate shows, after overnight incubation at 37°C, the characteristic colonies of klebsiella species; they are lightly pink in colour (due to lactose fermentation) and mucoid due to the production of extra-cellular slime. This latter feature also dictates the coalescence of adjacent colonies.

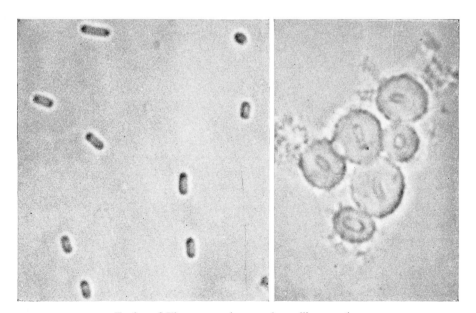

Typing of *Kl. aerogenes* by capsule swelling reaction.

On the left is a wet film preparation of *Klebsiella aerogenes*, type 54, mixed with *heterologous* antiserum: the *large* capsules, which were seen in parallel India ink films, are invisible. On the right is a wet film of the same organism mixed with its *homologous* antiserum and not only are the capsules 'swollen' but the bacilli are agglutinated. ×3750.

101

INFECTIONS CAUSED BY ENTEROBACTERIA

Whilst some species of the *Enterobacteriaceae* cause certain well-defined intestinal syndromes others are entirely commensal in the gut but are associated with infection in other tracts and tissues.

INTESTINAL INFECTIONS

In Britain intestinal infections caused by enterobacteria are bacillary dysentery, salmonella food poisoning, the enteric fevers and also gastroenteritis caused by enteropathogenic *Esch. coli*.

Bacillary dysentery occurs in all countries and has increased in incidence in Britain during the last 30-40 years so that it is now the commonest endemic bacterial infection in this country. The severity of bacillary dysentery is, in general, related to the different groups of shigellae and the more severe illnesses are usually caused by members of the *Sh. dysenteriae* group, the clinical picture produced by the serotypes of *Sh. flexneri* and *Sh. boydii* is less severe and cases of dysentery caused by *Sh. sonnei* are frequently so mild that the infection amounts to little more than a social inconvenience and the individual can continue at work.

The fact that many if not most cases of Sonne dysentery remain ambulant is very probably associated with the increasing incidence of the disease in Britain since such individuals have very much greater opportunities to spread the disease than those who are confined to their own homes.

Spread is essentially by the hand-to-mouth method, the case contaminates toilet fixtures, including wash basin taps and door handles, etc., and the organisms are thus transmitted to the hands and then to the mouths of other people; very occasionally Sonne dysentery may be water-borne when in rural areas a sewage outfall contaminates a river from which water is drawn for culinary and other human uses and is drunk without being efficiently purified.

Prevention, therefore, depends essentially on education of the individual in methods of maintaining a high standard of personal hygiene and since the infection is most common in school age groups there is also a need for the provision of *adequate* toilet facilities in schools; in particular, wash basins should be in the same cubicle as the toilet pedestal so that there is no need for communicating doors which must be handled before the user can wash his hands.

Salmonella food poisoning. As has already been stated, salmonella food poisoning is caused by members of the genus *Salmonella*, other than the four serotypes which cause the enteric fevers. As in bacillary dysentery infection is by ingestion, however, unlike dysentery, human cases and carriers are *not* the sole sources of infection; domestic animals, e.g., cows, sheep and pigs also act as sources and these also may be cases of infection or carriers. Rodents are heavily parasitised with salmonellae and can contaminate foodstuffs.

Meat intended for human consumption may be infected with salmonellae 'on the hoof' or it may originally have been free of salmonellae and become contaminated by various means during its preparation and serving, e.g., carcases can contaminate each other in the abattoir, during transportation to wholesale and then retail butcher's premises and there, clean meat can be contaminated via knives, choppers, etc., unless these are carefully cleaned between carcases. Similarly the human case or carrier involved in dressing the carcase can contaminate the meat and such human sources can also act when the carcase is being reduced to saleable fractions and in the ultimate stages of preparation at home or in kitchens for the preparation of communal meals, e.g., school-meal kitchens, hotels, etc.

Cow's milk and products derived from it may be a source of infection; in particular artificial cream is frequently incriminated in epidemics and this material is an excellent culture medium for many organisms including salmonellae so that foodstuffs incorporating contaminated artificial cream and stored at normal room temperature for hours or days before consumption will have a much higher population of salmonellae than if they had been eaten immediately after preparation.

Bird's carcases, especially those of hens, ducks and turkeys, can also act as a source of infection for man since they suffer from salmonella infection; and eggs may be the source of infection. Eggs may be contaminated from cloacal discharge during the cooling and drying period immediately after they are laid, or in the case of duck eggs infection may occur in the oviduct whilst the egg is being formed. Infection arising from eggs will involve only the individual consumer unless eggs are pooled in the preparation of a communal food, e.g., custards, when eggs which are free from infection may be contaminated if one infected egg is in the pooled material.

Obviously the prevention of salmonella food poisoning demands careful supervision of meat and other materials intended for human consumption at all levels of preparation from 'the hoof to the home'. Rodent proofing of premises prevents a further source of contamination and food which has been prepared but is not immediately consumed must be adequately stored, preferably in a refrigerator, so that any salmonellae which may be present do not have an opportunity to multiply.

The health of food-handlers should be constantly supervised and they must be educated and trained in methods of handling food which minimise or eliminate the risk of contamination.

The enteric fevers i.e., typhoid and paratyphoid fever, are spread predominantly by water supplies which are inadequately treated or stored and are contaminated from sources of infection, i.e., human cases and carriers. On occasion the role of water supplies may be an indirect one as in two recent epidemics of typhoid fever in Britain where the vehicle of infection was canned meat; the meat had, however, been contaminated originally from water containing *S. typhi*, used in processing the cans.

In Britain, *S. paratyphi B* is most frequently incriminated in epidemics of enteric fever and outbreaks due to *S. typhi* and *S. paratyphi A* are uncommon and in the case of *S. paratyphi C* virtually unknown; in communities with sophisticated water supplies and other environmental services and where cases of enteric fever are readily recognised and treated in isolation, carriers of enteric fever salmonellae are more likely to act as sources than are cases of infection. There are few foodstuffs which have not been incriminated as vehicles of infection and whilst some of these are imported in a contaminated state, e.g., desiccated coconut, other foods are contaminated by the hands of carriers employed in the food industry.

Therefore prevention depends on the institution and maintenance of safe water supplies and adequate methods of sewage disposal. People employed in water works, milk production—including those on dairy farms—and in the production of foodstuffs should be screened to ensure that they are not enteric carriers.

Reasonable protection can be provided by actively immunising individuals with *phenolised* TAB vaccine; such immunity should be offered to people living in or visiting countries with inadequate water supplies or sewage disposal systems but is not recommended for those living in well-developed communities.

Gastroenteritis caused by enteropathogenic strains of *Esch. coli* usually affects children in the first year of life and epidemics occur in residential nurseries and other institutions; the causal role of such strains of *Esch. coli* was recognised only some 25 years ago and babies who are artificially fed are much more likely to suffer infection than those who are breast-fed. Sources of infection are cases or carriers and spread is facilitated by fomites, particularly milk feeds which are not sterilised after being prepared and bottled; such *terminal sterilisation* eliminates any danger of infection by this method.

In the last few years there has been a definite decline in the reported incidence of gastroenteritis caused by enteropathogenic *Esch. coli*.

NON-INTESTINAL INFECTIONS

Non-intestinal infections caused by enterobacteria usually involve species other than those mentioned above in association with specific clinical syndromes; *urinary tract infections* with *Esch. coli.*, *Proteus* species, etc., may have an endogenous origin or alternatively exogenous infection may be acquired when the organisms are introduced as a result of instrumentation required for diagnostic or therapeutic reasons. Instruments may not have been properly sterilised prior to use or the preoperative cleansing of the patient may have been inadequate so that *Esch. coli* and other organisms present in the ano-genital region are carried into the bladder by the catheter or cystoscope.

Other non-intestinal infections caused by enterobacteria include infections of wounds, burns and bedsores and in such cases infection can originate by contact spread or dustborne spread from some other case or a carrier.

Certain serotypes of *Klebsiellae* are associated with a small proportion of cases of lobar pneumonia but such cases carry a high fatality rate; such types (serotypes 1 and 2) are given the specific name of *Kl. pneumoniae* (Friedlanders' bacillus).

CHAPTER 14

PSEUDOMONAS PYOCYANEA

THIS is the only member of the Pseudomonas group which is pathogenic to man. *Ps. pyocyanea* (*Ps. aeruginosa*) occurs widely in soil, sewage, water, etc.; it is commensal in the human intestine in small numbers and is also found, usually in association with other bacteria, in infected wounds or burns and urinary tract infections.

Microscopy. Identical to *Proteus* species but flagella are few in number and polar in origin.

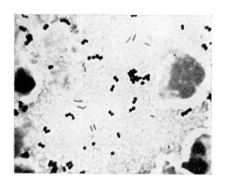

Gram-stained film of urinary deposit; culture of the specimen revealed that the Gram-negative bacilli were *Ps. pyocyanea* and the Gram-positive cocci were *Staph. pyogenes* var. *aureus.* ×1000.

Cultural appearances. Aerobic. Wide temperature range. Growth on nutrient agar gives colonies of 2 to 4 mm. in diameter which are convex and with an entire edge. Two striking characteristics are the sweet musty odour and the green coloration which diffuses into the medium.

Nutrient agar sensitivity test plate sown with a culture of *Ps. pyocyanea*; the natural colour of the medium (light straw) has been altered by pyocyanin pigment produced by the organism. Characteristically this strain shows resistance to many antibiotics, and indeed, in this case, is sensitive only to colomycin sulphate. ×⅔.

Biochemical activities. Of the sugars commonly employed in diagnostic laboratories only glucose is fermented and without gas production. Strains are encountered which do not produce the pigments characteristic of the majority; such cultures can be classified as *Ps. pyocyanea* if they give a positive oxidase reaction (see Chapter 8). Very few other Gram-negative bacilli are responsive to the oxidase reagent and these, in any case, react more slowly (more than 2 min.) than *Ps. pyocyanea* which responds within 30 sec.

Serological characteristics. Strains possess somatic and flagellar antigens but no valid serotyping scheme has been introduced; however, pyocine typing allows strains to be differentiated.

Pyocine typing. A method of typing strains for epidemiological tracing purposes has been developed recently and is now used in many countries; the technique is analogous to that of colicine typing of *Sh. sonnei* (pp. 94 and 95) except that incubation of the strain to be typed, i.e., the producer strain, must be at 32°C for 14-18 hr otherwise aberrant results will be given by certain strains. Similarly the eight indicator or passive strains are different from those used in colicine typing and all are *Ps. pyocyanea*; at least 37 different types within the genus can be recognised by their various patterns of inhibition on the indicator strains.

The patterns of inhibition illustrated here are those of pyocine type 1 and type 3 strains and these are the types most commonly found not only in Britain but in many other countries.

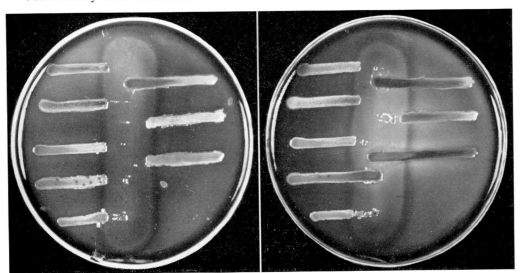

These plates were originally inoculated in the vertical plane with strains isolated from patients in an Assisted Respiration Unit; the plates were then incubated at 32°C for 14 hr and after the producer strain had been removed and the surface of the plate exposed to chloroform vapour the eight indicator strains were streaked on to the plate at right angles to the original growth line of the producer strains. The plates were then reincubated at 37°C for 18 hr. Indicator strains 1 to 5 are on the left of each plate, indicator strain 1 being at the top and similarly indicator strains 6 to 8 were streaked out on the right-hand side of each plate.
The type 1 producer strain has inhibited all of the indicator strains with the exception of strain no. 6; in the case of the type 3 producer strain indicator strains nos. 4, 6 and 8 have escaped inhibition.

INFECTIONS CAUSED BY PS. PYOCYANEA

In recent years Gram-negative bacilli have become increasingly significant as causes of hospital-acquired infection and *Ps. pyocyanea* has a special importance because of its resistance to most antibiotics.

Ps. pyocyanea is particularly dangerous when it infects debilitated patients, e.g., those who have suffered multiple injuries or individuals receiving radiation therapy; in such people infection frequently extends from a local site to become septicaemic.

Infection of the urinary tract, which is almost invariably exogenous, is common in patients on continuous drainage, e.g., paraplegic patients, and is often refractory to antibiotic therapy; the ability of *Ps. pyocyanea* to survive and indeed to grow in many disinfectants and antiseptics explains occasional epidemics in hospital practice, e.g., the use of multidose containers to administer eye-drops to patients who have undergone ophthalmic surgery has led to several recent epidemics of infection.

Another group of patients who are peculiarly susceptible to infection are those who require tracheostomy; it is essential that such individuals should be treated in strict isolation from each other and that respirators and other apparatus must be properly sterilised before being used for other patients.

CHAPTER 15

VIBRIOS AND SPIRILLA

THE genus *Vibrio* comprises members pathogenic to man, others pathogenic for animals or insects and many which are commensal or saprophytic particularly in water.

Popularly known as the comma bacillus and is the causative organism of Asiatic cholera.

Microscopy. When freshly isolated they appear as definitely curved rods with rounded ends, 3μ by 0.5μ. Often lose their curved appearance on artificial cultivation. Gram-negative. Motile with a single terminal flagellum. Non-capsulate and non-sporing.

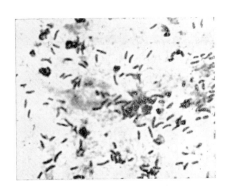

Gram-stained film of peptone water culture inoculated 6 hr previously with a piece of mucus from the stool of a cholera patient; numerous Gram-negative rods are present some of which have a typical vibrio appearance whereas others are straight. Cultivation on solid media revealed a mixture of *V. cholerae* and *Esch. coli.* ×1000.

Cultural appearances. Aerobic. Wide temperature range. Optimum 37°C. Grows on ordinary media but sensitive to acid pH; growth favoured by alkaline media and the optimum reaction is about pH 8·2. On agar the colonies are not distinctive, 2 to 3 mm. in size after 18 to 24 hr at 37°C, low convex with an entire edge, whitish and translucent. Older colonies develop a light ochre tint. Monsur's and Aronson's media are commonly employed for selective cultivation; by inoculating a tube of peptone water with a flake of mucus from the stool and incubating for only 6 to 8 hr vibrios, if present, can be harvested from the surface in almost pure culture.

Biochemical abilities. *V. cholerae* ferments glucose, sucrose, mannitol and maltose without gas production but does not utilise lactose or dulcitol. It gives a positive *Cholera-red reaction* when growing in peptone water due to the production of indole *and nitrites*. This can be tested by adding four drops of H_2SO_4 to a 72-hr culture.

V. cholerae is non-haemolytic when 1 ml. of a 2-day broth culture is added to 1 ml. of a 5 per cent. suspension of sheep red cells (Greig test).

Serological characters. There are three recognised serotypes of *V. cholerae* —'Inaba', 'Ogawa' and 'Hikojima'—all of which possess a common 'H' antigen but distinctive 'O' components.

EL TOR VIBRIO

This organism which has been incriminated in epidemic situations usually differs from the classical *V. cholerae* in being haemolytic in the Greig test; additionally, strains are resistant to certain phages and to polymyxin B (50 μg/disc) whereas *V. cholerae* strains are sensitive.

PARACHOLERA VIBRIOS

These vibrios, e.g., *Vibrio proteus*, are associated with choleraic illnesses which are less severe than true cholera. Some can be differentiated from *V. cholerae* by their biochemical reactions; most are haemolytic like the El Tor vibrio and although they possess an H antigen in common with *V. cholerae* they have specifically recognisable O antigens.

[CHOLERA

This disease which was once widespread throughout the world has now been contained in its original home in Asia; its absence from most other continents depends on the continued provision of filtered and chlorinated water which, after such treatment, is stored and delivered so that it cannot be contaminated.

Like typhoid fever, cholera is essentially a water-borne infection and the sources from which water supplies can be contaminated are cases of the disease; transient carriage can occur after recovery but carriage beyond a few weeks is rare.

Epidemic spread is frequently associated with the holding of religious and other festivals when the meagre sanitary arrangements break down and also in these circumstances case-to-case infection probably occurs via contaminated fomites and food.

Although El Tor vibrios can be distinguished from *V. cholerae* strains in the laboratory the clinical syndrome and the sources and modes of spread of cholera caused by the former organisms are identical with those of classical cholera.

Although active immunisation is practised there is no proof that this is protective for perhaps more than three to six months; emergency preventive measures involve temporary 'chlorination' of water supplies and the boiling of all water to be used for domestic purposes. On a long-term basis sanitary engineering has more to offer than medicine in the conquest of cholera.

Spirilla

Only one member of the genus *Spirillum* is pathogenic to animals and man. *Spirillum minus* is a rigid spiralled organism 3 to 5μ by 0·5μ. Gram-negative. Motile by virtue of bipolar lophotrichous flagella. Non-capsulate and non-sporing. Has not been cultured *in vitro*. Occurs in the blood and other tissues of rats and mice and may be transmitted to man if bitten by these rodents. This form of rat bite fever is uncommon in Europe but occurs in the Far East; the organism may be demonstrated in the local lesion, regional lymph glands and occasionally in the blood by microscopic examination of stained films or by intraperitoneal injection of such material into guinea-pigs which suffer septicaemia and death.

CHAPTER 16

HAEMOPHILI

MEMBERS of this genus are strictly parasitic in man or other animals and are feebly viable outside their respective hosts and, for growth require to be supplied with certain constituents of blood (haemophilic). *Haemophilus influenzae* is the commonest member of the genus pathogenic to man with *Haemophilus ducreyi* less commonly found.

HAEMOPHILUS INFLUENZAE

This organism is frequently found in the healthy human throat and is also associated with infection of the respiratory tract. It is the cause of a small proportion of cases of acute pyogenic meningitis in young children.

Microscopy. Characteristically small cocco-bacilli 1.5μ by 0.5μ. Gram-negative, non-motile, capsulate in young cultures, non-sporing. Often presents as very long filaments, particularly in cerebrospinal fluid from cases of haemophilus meningitis and also in old laboratory cultures.

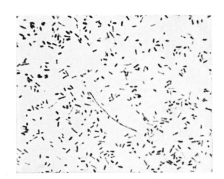
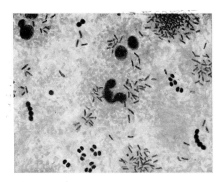

On the left is a Gram-stained film of a culture of *H. influenzae*; coccal and cocco-bacillary forms predominate but a few are bacillary and one filamentous form is evident. On the right, a Gram-stained film of sputum from a chronic bronchitic showing a multitude of *H. influenzae*, a few *N. catarrhalis* and *Strept. viridans*. ×1000.

Cultural appearances. Aerobic and demands blood-containing media for growth; blood agar or better, chocolate blood agar, provides the X (haematin) and V (diphosphopyridine nucleotide) factors required by *H. influenzae*. On these media, colonies are minute (1 to 2 mm.) and transparent; the organism grows in symbiosis with staphylococci which produce V factor. Hence in the vicinity of staphylococci, colonies of *H. influenzae* are larger in size—'Satellitism'.

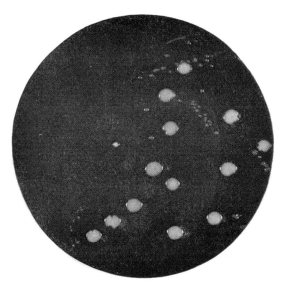

Satellitism: Small transparent colonies of *H. influenzae* along with larger pigmented colonies of *Staph. pyogenes* var. *aureus*; at the upper part of the plate a streak of *H. influenzae* colonies is barely visible but in the vicinity of the staphylococcal colonies the former are larger.

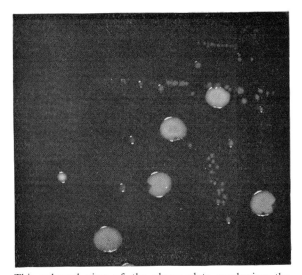

This enlarged view of the above plate emphasises the symbiotic relationship of these two species and also shows that regardless of their proximity to staphylococci, colonies of *H. influenzae* are larger in any case when widely separated from each other and thus get a greater share of the available nutrients.

115

Biochemical activities. Powers of fermenting carbohydrates are feeble and irregular.

Serological characters. Smooth strains possess capsular polysaccharide antigens; six types (a to f) can be recognised by a capsule swelling reaction analogous to that employed in typing pneumococci.

Animal inoculation. Not pathogenic for laboratory animals.

HAEMOPHILUS AEGYPTIUS

This species is for all practical purposes identical with *H. influenzae* and is associated with an acute and readily communicable type of conjunctivitis.

HAEMOPHILUS PARAINFLUENZAE

Such strains are morphologically similar to *H. influenzae* but require only the V factor for growth; they have more consistent and wider fermentative activities than *H. influenzae* and are associated with acute pharyngitis, acute conjunctivitis and rarely with ulcerative endocarditis.

HAEMOPHILUS DUCREYI

The causative organism of a venereally transmitted infection—chancroid or soft sore. In films made from the lesion or material aspirated from the swollen regional lymph glands, Gram-negative cocco-bacilli are noted and some may be present intracellularly.

Difficult to grow although requiring only the X factor and colonies resemble those of *H. influenzae*; agglutination tests with a specific antiserum confirm the identity.

INFECTIONS CAUSED BY H. INFLUENZAE

H. influenzae can be isolated from the healthy human throat in 40-70 per cent. of people and appears to fulfil a commensal role in most instances; the majority of such strains are rough forms.

For many years influenza bacilli were regarded, and wrongly, as the cause of influenza but its role as a secondary bacterial pathogen in this virus infection is now accepted; this is the usual feature in infections involving *H. influenzae*, namely that it is found in association with other organisms in respiratory tract infections, e.g., with pneumococci in sinusitis, bronchitis and bronchiectasis.

There would appear to be little doubt, however, that this organism is actively involved in the disease process since its elimination by suitable antibiotic therapy is accompanied by an improvement in the clinical state; similarly long-term prophylactic administration of antibiotics to those suffering from chronic bronchitis prevents acute episodes in which *H. influenzae* is commonly found in the sputum.

The unusual morphology of *H. influenzae* in films of CSF from cases of haemophilus meningitis has already been noted and in this condition the organism acts as the primary pathogen; haemophilus meningitis is most common in children of pre-school age.

CHAPTER 17

BORDETELLAE

THIS generic title commemorates the isolation by Bordet of the whooping-cough bacillus, *Bordetella pertussis*; this organism, along with two closely related species *Bord. parapertussis* and *Bord. bronchiseptica*, was originally classified with *Haemophilus* but since they require neither X nor V factors for growth they receive separate status.

BORDETELLA PERTUSSIS

Microscopy. Bears a close resemblance to *H. influenzae* but is less pleomorphic.

Cultural appearances. Complex enriched media, such as Bordet-Gengou's potato blood glycerol agar, are required for primary isolation; even then, 2 to 3 days incubation at 37°C are required before colonies can be recognised. These are small, dome-shaped and highly refractile.

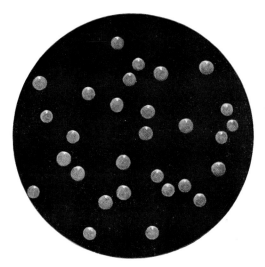

Colonies of *Bord. pertussis* on Bordet-Gengou medium
after 4 days incubation at 37°C. ×5.

Biochemical activities. No reliable fermentative properties.

Serological characters. Freshly isolated strains are serologically homogeneous—phase 1 organisms.

BORDETELLA PARAPERTUSSIS

This organism has been isolated from a whooping-cough-like disease. Similar in many respects to *Bord. pertussis* but produces darkening of underlying blood in Bordet-Gengou medium. Unlike *Bord. pertussis* it produces urease. Strains are serologically homogeneous and although there are antigenic components in common with *Bord. pertussis, parapertussis* strains can be specifically agglutinated with an absorbed antiserum.

BORDETELLA BRONCHISEPTICA

Rarely associated with man, but causes bronchopneumonia in rodents and has been isolated from dogs suffering from distemper. Morphologically it differs from the other members of the genus in possessing peritrichous flagella. Grows readily on ordinary media without blood and colonies are very similar to those of *Bord. pertussis*. Urease is produced. Serologically related to other members of the genus but capable of differentiation by employing absorbed antiserum.

WHOOPING-COUGH

Whooping-cough or pertussis is essentially a disease of childhood and although the fatality rate is low, approximately 50 per cent. of deaths occur in those infected in the first six months of life.

It is probable that infection is spread by direct droplet spray from an infected individual but the relative importance of the part played by other methods of spread such as indirect contact via fomites and dust-borne spread is not yet known.

The disease is most highly communicable in the early catarrhal stage before the typical paroxysmal cough or whoop appears, thereafter communicability decreases fairly rapidly so that within two to three weeks after the onset of whooping the case is no longer a source of infection, although coughing may continue for several more weeks.

Whooping-cough is one of the few infections which occur more frequently and are more severe in females than in males; in communities unprotected by active immunisation 80 per cent. or more of the children suffer infection and the disease recurs in epidemic fashion every three to four years.

Active immunisation with a potent pertussis vaccine should be undertaken in the first six months of life beginning at the second month; in this way we can influence the incidence of the disease at a time when the fatality rate is highest.

In comparison with a 90 per cent. attack rate among unimmunised siblings, the attack rate for protected children with home exposure is, with a potent vaccine, less than 10 per cent.

Whooping-cough does occur in immunised children, but the disease is very much milder and of much shorter duration than in unprotected children.

Film of conjunctival exudate from a case of conjuncti-vitis due to *Moraxella lacunata*; a short, thick, Gram-negative diplo-bacillus. This organism was previously classified with the haemophili but does not require accessory growth factors; it grows readily on serum agar or blood agar and its specific name is associated with its ability to liquefy such media so that the colonies float on fluid in lacunae or craters in the solid medium. ×1000.

CHAPTER 18

PASTEURELLAE

THE only member of this genus which is readily transmissible to man is *Pasteurella pestis* whose natural hosts are rats and certain other rodents, whence it is transmitted by rat fleas, e.g., *Xenopsylla cheopis*. Other pasteurellae are incriminated in haemorrhagic septicaemias in various animals and birds.

PASTEURELLA PESTIS

The cause of human plague which has, in the past, decimated the world population from time to time.

Microscopy. Short ovoid bacilli, $1·5\mu$ by $0·7\mu$, Gram-negative, non-motile, capsulate in tissues or when freshly isolated. Non-sporing. Characteristically shows bipolar staining. Pleomorphic on prolonged cultivation or subculture.

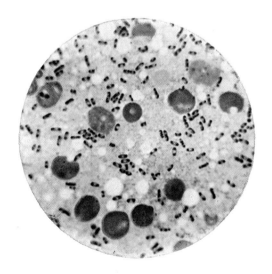

Smear from enlarged lymph gland from a case of plague; numerous *P. pestis* are seen as ovoid bacilli with characteristic bipolar staining. Film treated with Leishman's stain. ×1000.

Cultural appearances. Aerobic and facultatively anaerobic. Optimum temperature=30°C. Grows on ordinary nutrient agar but more rapidly if blood or serum is added. Colonies are small (1 mm.), greyish and semi-transparent but become larger and irregular in outline on continued cultivation.

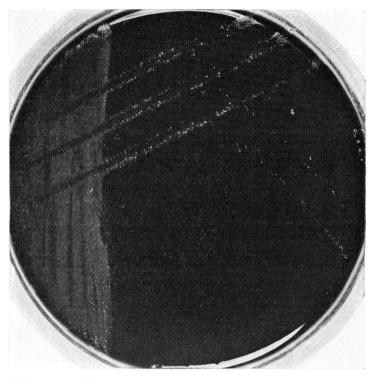

48 hr culture of *P. pestis* incubated at 30°C. Colonies are small, greyish and semitransparent; this particular strain, unlike the majority, could not be grown on MacConkey's medium in spite of being biochemically and serologically proven to be *P. pestis*.

Biochemical activities. These are shown in the Table but there is some variation in strains within the species.

Serological characters. Strains are serologically homogeneous.

Animal inoculation. Guinea-pigs, rats and other rodents are susceptible. The subcutaneous injection of a culture into a white rat causes death within a few days. Post-mortem reveals marked local oedema and necrosis; the related lymph glands are markedly enlarged and the spleen congested and may show small white areas in its substance. The bacilli are recoverable from these sites and also from the heart-blood in the terminal stages of the illness.

Differential characteristics of some Pasteurellae

	Motility at 22°C.	Maltose	Indole	Growth on MacConkey's Medium
P. pestis . . .	—	⊥	—	+Scanty
P. tularensis . .	—	⊥	—	—
P. septica . . .	—	—	+	—
P. haemolytica . .	—	⊥	—	—
P. pseudotuberculosis .	+	⊥	—	+Abundant

PASTEURELLA TULARENSIS

The causative organism of tularaemia in wild rodents and transmissible to man by fleas and ticks, e.g., as an occupational hazard in persons handling rabbits. Originally reported from Tulare County, California.

Microscopy. Much smaller ($0·7\mu$ by $0·2\mu$) than *P. pestis* but otherwise similar and shows bipolar staining.

Cultural appearances. Aerobe. Will not grow unless complex media are provided, e.g., blood agar containing added cystine and glucose. Optimum temperature 37°C. Colonies similar to those of *P. pestis* and often showing α-haemolysis.

Biochemical activities. Feebly fermentative without gas production, in various sugars.

Serological characters. So far as is known strains are relatively homogeneous but show some extra-generic relationships, particularly with brucellae.

Animal inoculation. Most rodents can be infected experimentally and the organism is recoverable from heart-blood, liver and spleen.

The other members of the genus *Pasteurella* are rarely pathogenic to man, although *P. pseudotuberculosis* has been incriminated in some cases of acute mesenteric lymphadenitis and *P. septica* has been recovered from wounds inflicted by the bite of dogs or cats and its presence is thought to delay wound healing.

PASTEURELLA SEPTICA

P. septica strains affect various animals and are usually named according to their animal host, e.g., *P. boviseptica* in cattle, but are possibly members of the same species differing only in their parasitic adaptation.

PASTEURELLA HAEMOLYTICA

P. haemolytica causes pneumonia in cattle and sheep and also one type of septicaemia in lambs.

PASTEURELLA PSEUDOTUBERCULOSIS

P. pseudotuberculosis causes a tuberculosis-like infection in guinea-pigs and certain other rodents. Unlike other members of the genus it is motile at 22°C.

PLAGUE

Plague is essentially an epizootic infection in wild rats and certain other rodents; man is affected only in circumstances where his environment allows the vector rat flea easy access to him, e.g., overcrowding in insanitary living quarters.

Bubonic plague is the type most commonly seen in man and results from the bite of an infective rat flea; the rat flea becomes infected by taking a blood meal from a rat and the bacilli then multiply in the stomach and proventriculus. Then, if and when the flea attempts to take a blood meal from man plague bacilli are regurgitated from the proventriculus into the tissues.

A few days after being inoculated by the flea the patient shows progressive swelling of the regional lymph glands and surrounding tissues—this lesion is known as the primary bubo; secondary buboes may occur in other lymph glands. The buboes are packed with plague bacilli.

In bubonic plague infection rarely spreads from person to person but if a septicaemic phase develops then a rapidly fatal bronchopneumonia occurs and this *pneumonic plague* can spread to other healthy people by airborne routes and since *P. pestis* can survive for several weeks on sputum-contaminated surfaces, pneumonic plague may be dust-borne as well as being spread by droplet nuclei.

Prevention of plague depends on rodent control and destruction of fleas; during epidemics individuals at special risk must wear protective clothing and masks. Medical and nursing personnel should be given prophylactic doses of broad-spectrum antibiotics.

CHAPTER 19

BRUCELLAE

ALL members of this genus are strictly parasitic on man or animals and are characteristically intracellular. Brucellosis in the human subject is classically associated with drinking unpasteurised goat's milk (*Brucella melitensis*) or cow's milk (*Brucella abortus*). In either case and also with *Brucella suis* (affecting pigs) there is also an occupational hazard to farm workers, abattoir personnel and veterinary surgeons. Laboratory workers may become infected from handling cultures.

BRUCELLA MELITENSIS

Microscopy. Varies in shape from coccal forms (0·5μ) to small bacilli (1μ by 0.5μ). Gram-negative. Non-motile. Capsulate. Non-sporing.

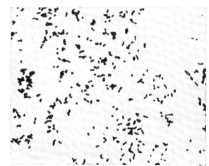

Gram-stained film of *Br. melitensis* isolated in Britain from a person recently returned from the Middle East and who had, for a few weeks, been addicted to drinking raw goat's milk. Gram-negative coccal and cocco-bacillary forms are seen. ×1000.

Cultural appearances. Aerobe. Grows on ordinary media but a more rapid and luxuriant growth is obtained by cultivation on liver-infusion agar. Colonies are low convex, with an entire edge, transparent and approximately 1 mm. in diameter; on continued incubation they increase in size and become brownish.

Colonies of *Br. melitensis* isolated on blood agar after 48 hr incubation at 37°C. Culture was from the case mentioned above and has the features noted in the above paragraph. ×3

Biochemical abilities. In ordinary sugar media no fermentation is observable but if a peptone-free, buffered medium containing the relevant substrate is heavily inoculated very consistent patterns are obtained.

Serological characters. *Br. melitensis* is very closely related to the other members of the genus since all possess two similar antigens; however, one of these latter is dominant in *Br. melitensis*, thus it is possible to prepare an absorbed agglutinating antiserum for this species.

Animal inoculation. Guinea-pigs inoculated intramuscularly or subcutaneously with *Br. melitensis* suffer a chronic illness which is rarely fatal and differs from that caused by inoculation with other brucellae.

BRUCELLA ABORTUS: BRUCELLA SUIS

These are closely similar to *Br. melitensis* in most biological characteristics. Differential characteristics are shown below.

Differential characteristics of Brucellae

	Br. melitensis	*Br. abortus*	*Br. suis**
Glucose.	⊥	⊥	⊥
Inositol	—	⊥	—
Maltose.	—	—	⊥
Inhibited by:			
a. Basic Fuchsin 1:25,000 .	No	No	Yes
b. Thionine 1:30,000 . .	No	Yes	No
CO_2 required for growth .	No	Yes	No
H_2S produced . . .	No	Yes	Yes

* American strains. Danish strains are similar except for inability to produce H_2S.

BRUCELLOSIS

Brucellosis is primarily an infection of certain animals, and in Britain *Br. abortus* is the species which is most commonly incriminated in bovine and human infections.

Whilst the general population is at risk of infection if unpasteurised milk from an infected cow is drunk, there are certain occupations with a high risk of acquiring infection by coming in contact with sick animals or their discharges. Similarly the handling of infected carcases allows the organisms to pass through skin abrasions or to be inhaled; abortion is a common phenomenon in infected cattle but is rare in infected pregnant women and it is thought that this is explained by the fact that whereas bovine and other animal placentae contain erythritol which stimulates the growth of brucellae, erythritol is absent from the human placenta.

The widespread immunisation of cattle has cut dramatically the incidence of abortion but infection in cattle is still endemic despite immunisation.

The severity and duration of brucellosis in man are subject to wide variation and the disease should be considered as a possible cause when cases of pyrexia of uncertain origin are being investigated. As with almost all other zoonoses, infection rarely spreads from man to man.

Prevention of brucellosis in those at occupational risk demands the education of farmers, abattoir workers and others handling carcases in methods of minimising the risk and particularly in regard to the manipulation and disposal of discharges and foetuses from aborted animals.

Whilst the prevention of brucellosis in the general population is dependent on the eradication of the disease in the animal population, pasteurisation of all milk offers an effective safeguard against acquiring infection by this vehicle.

CHAPTER 20

BORRELIAE

THE members of this genus are pathogenic or potentially pathogenic for man or animals. As a genus they differ from other spirochaetes in being larger and therefore visible with the compound light microscope after staining with normal dyes.

BORRELIA VINCENTII

This is the only member of the genus found in Britain; it leads a commensal existence in the healthy human mouth but when the mucous membrane is devitalised, e.g., by trauma or nutritional deficiency, *Borr. vincentii* may assume a pathogenic role in association with fusiform (cigar-shaped) bacilli. Vincent's infection is frequently seen as an ulcerative gingivo-stomatitis and occasionally as ulcerative tonsillitis. These symbiotic organisms are sometimes found in lung abscesses and cases of bronchiectasis.

The diagnosis of Vincent's infection depends on the microscopic findings of *large numbers* of fusiform bacilli and *Borr. vincentii*; the latter is 5 to 20μ by 0·3μ and possesses three to eight irregular spirals; Gram-negative and non-sporing. It is strictly anaerobic and very difficult to cultivate.

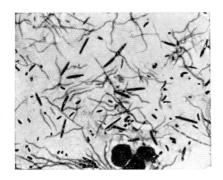

Film from a case of gingivo-stomatitis after staining for 10 min. with dilute (1 : 10) carbol fuchsin. **Large** numbers of *Borr. vincentii* and fusiform bacilli (*Fusiformis fusiformis*) are seen. Our inability to grow these species easily demands that, at present, we rely on such microscopic appearances for diagnosis. Vincent's infection responds rapidly to penicillin therapy. ×1000.

BORRELIA RECURRENTIS

The causative organism of European relapsing fever.

Microscopy. 10 to 30μ by 0·3μ with five to seven fairly regular coils. Gram-negative. Non-sporing.

Cultural appearances. The organism is very difficult to grow even in heavily enriched fluid media. Not grown on solid media.

Biochemical abilities. No knowledge available.

Serological characters. No detailed knowledge but antiserum developed against *Borr. recurrentis* does not agglutinate *Borr. duttonii* which is responsible for West African relapsing fever.

Animal inoculation. White mice, but not guinea-pigs, can be infected by subcutaneous inoculation of blood from a human case of relapsing fever; the organism can be demonstrated in stained films of tail blood 1 to 3 days later.

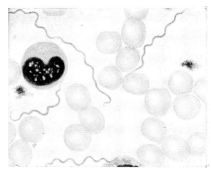

Blood film from a case of relapsing fever in Poland; four *Borr. recurrentis* can be seen, and that above the mononuclear leucocyte appears to be dividing. Morphologically identical with other species involved in relapsing fever in various parts of the world. Film treated with Leishman's stain. ×1000.

BORRELIA DUTTONII

Identical with *Borr. recurrentis* except for its serological characters; it is antigenically distinct from all other relapsing fever spirochaetes.

Many other Borreliae have been associated with relapsing fever in different parts of the world and specific names have been given to them often on flimsy evidence.

RELAPSING FEVERS

European relapsing fever is louse-borne from man to man by *Pediculus humanus* var. *corporis*; after the louse has taken a blood meal from an infected individual the spirochaetes can be demonstrated in the louse's stomach for 12 to 24 hr and then they disappear. Some 5 to 7 days later they reappear throughout the louse's body and man can then be infected either by crushing lice on his skin whilst scratching and thus introducing the spirochaetes through abrasions; alternatively the insect may bite the patient and the bite-wound can then be contaminated with the louse's infective excreta.

Prevention depends primarily on delousing of individuals, their clothing and the household environment.

West African relapsing fever is tick-borne and the natural reservoirs of infection in this type of relapsing fever are wild rodents and also the ticks themselves since the spirochaetes are transmitted transovarially to consecutive generations of ticks.

Obviously the control of tick-borne relapsing fever is more difficult than that of louse-borne infection since mammals other than man are involved in the cycle of infection and also, ticks once infected may remain infective for several years.

CHAPTER 21

TREPONEMATA

THIS is a large genus comprising a few pathogenic members of which only one, *Treponema pallidum*, is found in Britain. Many of the commensal species occur in situations, e.g., genitalia where their differentiation from *Tr. pallidum* is of importance in the diagnosis of syphilis.

TREPONEMA PALLIDUM

The causative organism of syphilis.

Microscopy. Delicate spiralled filament 6 to 14μ by 0·1μ with 6 to 12 small regular coils. Ends are pointed and tapering and never recurved. Feebly refractile and requires dark-ground illumination for its demonstration in the unstained state. Cannot be demonstrated by ordinary staining techniques; silver impregnation stains may be used.

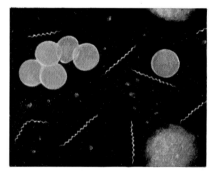

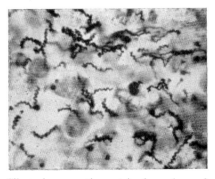

Dark-ground illumination preparation of serous exudate from primary syphilitic chancre. Seven delicately spiralled *Tr. pallidum* are seen, that on the left showing flexion. Red blood cells and part of two leucocytes are also seen in the field. ×1000.

Silver impregnation stained section of foetal liver from a case of congenital syphilis; the liver tissue is counterstained with basic fuchsin. The organisms are much rougher and thicker in appearance as compared with the dark-ground preparation and this is due to the deposition of silver. ×1000.

Cultural appearances. Has not been cultivated in artificial media and therefore nothing is known of its biochemical abilities.

Serological characters. The antigenic make-up is unknown but in man the organism stimulates formation of antibodies which can immobilise live motile *Tr. pallidum* and which give a positive complement-fixation test (Wassermann test) with a suspension of a phosphatide lipid extracted from normal animal tissues. Thus syphilis can be diagnosed serologically.

TREPONEMA PERTENUE

The cause of yaws, a non-venereal but communicable disease in tropical countries. Identical with *Tr. pallidum* but originally regarded as more slender, hence the specific name.

In Arabia, another non-venereal disease, bejel, and in Tropical America and certain areas of the Pacific, the disease pinta, are both apparently caused by organisms indistinguishable from *Tr. pallidum*; the specific name *Tr. carateum* is used to describe the organism associated with pinta.

In all of these non-venereal treponematoses the Wassermann test becomes positive.

COMMENSAL TREPONEMATA

These are found on the genitalia or in the mouth; *Tr. gracile* and *genitalis* in the former situation and *Tr. microdentium* in the latter. Microscopic differentiation from *Tr. pallidum* is difficult and must be undertaken with great care; these commensals can be grown in fluid media under strictly anaerobic conditions. These surface commensals can be avoided if great care is taken to prepare the area of the suspect syphilitic sore by thorough cleansing before collecting serous exudate.

SYPHILIS

With few exceptions syphilis is contracted by sexual intercourse; *Tr. pallidum* is so feebly viable outside the host's tissues that infection acquired other than by intercourse *always* involves direct contact, e.g., manual infection in doctors and nurses who have examined carelessly a syphilitic lesion.

The primary lesion or chancre appears 3 to 6 weeks after exposure to infection and is at first papular but necrosis occurs with the formation of an indurated ulcer and *Tr. pallidum* is present in large numbers in the primary chancre which heals with the formation of a scar.

Secondary stage lesions appear 6 to 12 weeks after the appearance of the chancre and during this stage the spirochaetes spread throughout the patient's tissues via the bloodstream; generalised skin rashes, condylomata of the anus and vulva and mucous patches in the mouth are frequently seen in the secondary stage and all of these lesions contain large numbers of the causal spirochaete but their inability to survive on clothing, etc., fortunately protects innocent contact with the patient.

Tertiary syphilis is characterised by the appearance of gummata in various organs some years after infection unless treatment has eliminated the infection; lesions in the central nervous system give rise to characteristic clinical syndromes, e.g., tabes dorsalis.

The control of sexually-acquired syphilis is similar to that of all venereal diseases and includes health and sex education, the provision of clinics for rapid diagnosis and treatment, the intensive follow-up of possible sources of infection and the screening and, if need be, treatment of contacts.

The ease with which gonorrhoea can be treated outside venereology departments carries the risk of masking concomitant primary syphilis, thus all cases of gonorrhoea should be kept under serological surveillance to ensure that syphilis was not acquired simultaneously.

Congenital syphilis has diminished in incidence in recent years and much of the reduction can be attributed to careful antenatal care of the pregnant woman which must include serological tests to exclude the possibility that she is suffering from syphilis; alternatively intensive antibiotic therapy of the syphilitic expectant mother may reduce the chances of her child suffering infection.

CHAPTER 22

LEPTOSPIRAE

MORE than 80 pathogenic members of the genus have so far been recognised but these vary in their geographic distribution and host range. Rats, dogs, field mice and pigs are natural hosts for some leptospirae and man may be parasitised from these sources. In Britain, only two species have so far been incriminated in human disease, *Leptospira icterohaemorrhagiae* and *Leptospira canicola*.

LEPTOSPIRA ICTEROHAEMORRHAGIAE

The causative organism of Weil's disease.

Microscopy. Finely coiled and measuring 6 to 20μ by 0·1μ. One or both ends of the organism are hooked or recurved on the body. Requires dark-ground illumination for its demonstration if unstained; silver impregnation stains are used to allow visualisation by normal illumination.

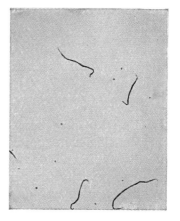

<table>
<tr>
<td>Dark-ground preparation of L. icterohaemorrhagiae from Stuart's medium. Both ends of the organism are recurved on the body; the very fine coils are just discernible. ×1000.</td>
<td>Silver impregnation stained preparation of L. canicola which is morphologically indistinguishable from other leptospires; in comparison with the living preparation on the left the leptospires here show a tendency to straighten and lose their characteristic hooked appearance. ×1000.</td>
</tr>
</table>

Cultural appearances. Requires special fluid media for growth, e.g., Stuart's medium which contains animal serum. The organism is particularly susceptible to pH below 7. Growth on solid media has not yet been introduced for diagnostic purposes. Optimum temperature=30°C.

Biochemical abilities. No knowledge available.

Serological characters. Because of the morphological similarity of pathogenic leptospirae and the absence of cultural or biochemical criteria for differentiation, their antigenic constitution is at present the most satisfactory method of
species identification for epidemiologic purposes. *L. icterohaemorrhagiae* is
related to many other species by virtue of 'group' antigens but an absorbed
antiserum allows type identification.

Animal inoculation. Young guinea-pigs are very susceptible to intraperitoneal inoculation either of cultures or of blood or urine specimens obtained
from a patient. The animal dies in 1 to 2 weeks with jaundice apparent on
serous membranes and haemorrhages in the lungs and muscles.

The ability of leptospirae to pass through skin is taken advantage of in
attempting to demonstrate their presence in water; a young guinea-pig, whose
belly has been shaved, is partially immersed in the water for 1 hr. If infection
takes place the animal, at post-mortem, displays the characteristics mentioned
above.

LEPTOSPIRA CANICOLA

This organism is morphologically identical to *L. icterohaemorrhagiae* and
can be differentiated only by serological tests. Its natural host is the dog which
often remains apparently healthy or may suffer nephritis with death. In either
case, humans may be infected from the dog's urine. Canicola fever in man is a
much milder form of leptospirosis than Weil's disease and rarely fatal.

SAPROPHYTIC LEPTOSPIRAE

These are frequently present in water either in nature or from domestic
taps. They are similar in appearance to pathogenic members of the genus but
are readily distinguished by growing in simple media without added serum and
in lacking pathogenicity for guinea-pigs. Such saprophytic species are collectively termed *Leptospira biflexa*.

LEPTOSPIROSIS

Although specific names have been given to the clinical illnesses resulting
from infection with leptospirae, e.g., canicola fever, Fort Bragg fever, seven-day
fever, etc., the epidemiology of leptospiral infection is identical regardless of the
particular serotype or natural host involved. The classical leptospiral infection
is known as Weil's disease and the natural host of the causal leptospire, *L.
icterohaemorrhagiae*, is the brown rat (*Rattus norvegicus*), which frequently
remains healthy and the leptospirae are excreted in the rat's urine. Provided the
urine is voided in moist, alkaline surroundings the spirochaetes may remain
viable for many days and if man comes into contact with them they can penetrate
skin and mucosal surfaces through cuts and abrasions.

Although infection may occur by bathing in polluted water most cases are associated with an occupational risk, i.e., occupations involving work in rat-infested places which are also moist, such as coal mines, fish-gutting halls or sewer systems. In similar fashion and in other countries workers in rice-fields or sugar-cane workers are at risk from other leptospires; different serotypes have adapted to different animal hosts and whilst these latter are usually rodents, *L. canicola* has as its host the dog and also the pig.

Control of leptospiral infection depends on elimination of the natural host, the modification of the environment in occupations with a high risk, e.g., rodent proofing of fish-gutting halls and the spraying of working surfaces in such premises with an acid solution at the beginning and end of each working period, the provision of protective clothing for sewer workers, etc.; when cases of leptospirosis are traced to a domestic animal then the animal must either be sacrificed or treated to eliminate leptospiral parasitisation.

Direct transmission from man to man is extremely rare.

CHAPTER 23

ACTINOMYCETES

THIS heterogeneous collection of microorganisms may superficially resemble fungi but are related to true bacteria; characteristically they form a branching mycelium which tends to fragment into coccal and bacillary pieces. Many actinomycetes are entirely saprophytic, particularly in soil; a few produce disease in man and animals.

ACTINOMYCES

This genus comprises obligately anaerobic or microaerophilic members which are parasitic on man and certain domestic animals. Two species may be recognised, *Actino. israelii* (human in origin) and *Actino. bovis* (from cattle and pigs) but they have many characteristics in common.

Microscopy. Filamentous branching organism, Gram-positive, non-motile, non-capsulate and non-sporing. In tissues from infected individuals 'sulphur granules' are formed which comprise a central mycelial mass with a peripheral zone of swollen clubs which stain Gram-negatively and are acid-fast provided that only 1 per cent. H_2SO_4 is used in attempted decolorisation. Club formation is more marked in bovine than in human lesions.

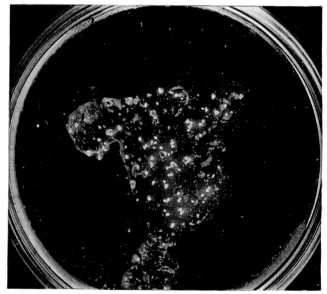

'Sulphur granules.' These granules were obtained from a pleural effusion in a case of thoracic actinomycosis with secondary staphylococcal infection. The granules vary in size and a few of the largest granules are showing a darker, almost brownish colour.

138

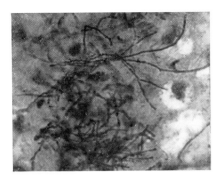 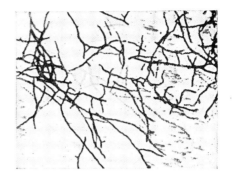

On the left is a Gram-stained preparation of pus from the case of thoracic actinomycosis referred to in the previous illustration; note the Gram-positive branching filaments with evidence of fragmentation into bacillary and coccal forms; *on the right* is a Gram-stained film made from an anaerobic culture plate which was inoculated with a 'sulphur granule' from the lesion.

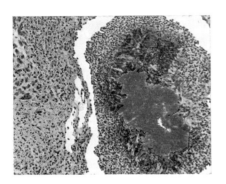 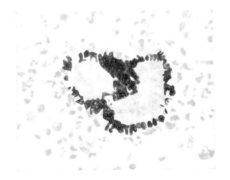

On the left is a Gram-stained section of human liver showing an actinomycotic abscess; *on the right* is a Ziehl-Neelsen stained preparation of an actinomycotic colony ('sulphur granule') from an ox-tongue; the acid-fast peripheral zone of clubs is clearly seen. ×75, ×500.

Cultural appearances. Primary isolation must be by anaerobic cultivation on blood agar or in a shake culture of serum agar in which colonies form in a band 10 to 20 mm. below the surface of the medium. When grown as surface colonies there is much variation in appearance but these usually present a rough appearance, irregular in outline, cream or white in colour and in the case of *Actino. israelii* are often firmly adherent to the medium.

Biochemical abilities. *Actino. bovis*, unlike *Actino. israelii*, hydrolyses starch but otherwise their fermentative abilities vary with different strains within each species and do not allow reliable differentiation.

Serological characters. Four groups can be recognised but, at present, such techniques are not employed in diagnosis.

Animal inoculation. Results are irregular and of no value in diagnosis.

ACTINOMYCOSIS

Actinomycosis in the human is an endogenous infection with *Actino. israelii* derived from the mouth where they live an essentially commensal existence; they can be isolated from tonsils and from carious teeth in 5 per cent. of healthy individuals.

The majority of cases of human actinomycosis occur in the cervico-orofacial region (65 per cent.) with abdominal actinomycosis being next most common (19 per cent.); there is almost always a history of trauma preceding the onset of infection, e.g., extraction of a carious tooth or an accidental blow.

The disease presents many interesting epidemiological problems and although the dissociation from bovine actinomycosis has been proven, male agricultural workers have a very much higher incidence of infection than men in other types of employment; similarly men suffer more frequently than women (3 to 1) and although the disease occurs at all ages, more than half of the incidence occurs between the ages of 10 and 29 years. Case to case infection is unknown.

SECTION III

DIAGNOSTIC METHODS

Some of the techniques of bacteriologic practice, and in particular the processing of specimens, are outlined in this section. In the processing charts the following abbreviations have been used to indicate various culture media:—

NA   = Nutrient agar.
BA   = Blood agar.
CVBA = Crystal violet blood agar.
M^{C}C   = MacConkey's medium.
DCA   = Desoxycholate citrate agar.
CMB   = Cooked-meat broth.
W & H = Willis & Hobb's medium.

Cross reference to relevant chapters in Section II should be made in studying the various processing charts.

In all instances, specimens being submitted for cultivation should be transmitted immediately after being taken; if this is impracticable they should be refrigerated. If the patient is to receive antibiotic therapy, specimens should be collected before treatment is started or if the individual is already under such treatment the bacteriologist must be informed of this on the accompanying request form. This form should state not only the patient's name, age, sex and address but a précis of the clinical history and an indication of the submission of previous specimens; the date and time of taking the specimen should also be noted on the request form.

CHAPTER 24

BLOOD CULTURE

BACTERAEMIA is a constant finding in many human infections and usually occurs in the early stages of illness; the attempted isolation of organisms from the blood is undertaken less frequently than it might be since the identification of an accepted pathogen from the blood stream offers a definite and probably early diagnosis, e.g., in the enteric fevers the causal salmonellae can be isolated from the blood very readily during the first week of the illness when isolations from the stool are less common.

Materials required at bedside.

(a) Soap and water, surgical spirit and iodine (2 per cent. in 70 per cent. alcohol) with swabs to apply these cleansing agents.

(b) Tourniquet to render the veins turgid in the ante-cubital fossa.

(c) 10 ml. all-glass syringe sterilised by dry heat (hot air oven 160°C/1 hr); with additional needle.

(d) Blood culture bottle; this contains 50 ml. of one of several media all of which incorporate 'Liquoid' (0·05 per cent.) which is the proprietory name for sodium polyanethyl sulphonate. 'Liquoid' acts as an anti-coagulant and also annuls the natural bactericidal action of blood.

The screw-cap top of the bottle is kept sterile by a 'viskap' and this cover can be removed by pulling upwards on the linen rip-cord to expose the sterile metal-cap with a central hole under which is a rubber diaphragm through which the blood is injected from the syringe. By this method the contents of the blood culture bottle are not exposed to the air during inoculation.

Bedside procedure. An assistant should be available to control the tourniquet, rip off the 'viskap' and attend to the patient's arm after venepuncture.

The ante-cubital area should be thoroughly washed with soap and water, then with surgical spirit and finally treated with iodine, in order to reduce as far as possible contamination of the specimen with skin organisms. The tourniquet is applied and if necessary the veins made more prominent by the patient clenching and opening his hand; the syringe is removed from its sterile tube without touching the needle and 5 to 10 ml. of blood are withdrawn from a vein. The tourniquet is then released, the syringe and needle withdrawn and pressure applied to the puncture area.

The needle is detached from the syringe and, using forceps, is replaced by a fresh sterile needle; the 'viskap' is ripped off the blood culture bottle and the rubber diaphragm pierced with the freshly attached sterile needle and the contents of the syringe injected into the bottle which is then agitated gently for a few seconds to ensure that the blood does not clot.

The syringe and needle are withdrawn from the diaphragm as one unit and the rubber diaphragm swabbed with surgical spirit.

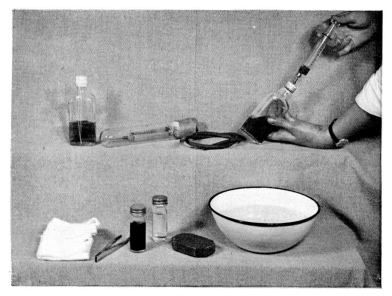

Materials required for blood culture

Reading from left to right, the materials in the top row are: 1. A blood culture bottle with 'viskap' intact. 2. A dry, sterilised 10 ml. all-glass syringe with needle fitted; the needle is sheathed by an open-ended piece of glass tubing which protects the point from mechanical damage against the container. 3. A rubber tourniquet. 4. A blood culture bottle, from which the 'viskap' has been removed by pulling on the linen rip-cord, into which blood is being introduced through the sterile rubber diaphragm.

In the bottom row: 1. Gauze swabs. 2. Iodine. 3. Surgical spirit. 4. Soap and water, all for use in cleansing the skin prior to venepuncture.

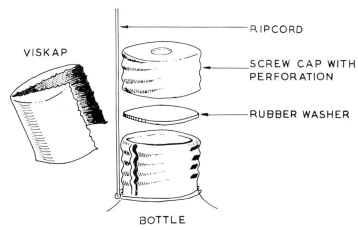

Exploded view of top of blood culture bottle.

Laboratory procedure. The inoculated bottle is returned as rapidly as possible to the laboratory where it is immediately placed in a 37°C incubator after having an identifying label attached.

After 18 to 24 hr incubation, two or three loopfuls of blood/broth mixture are removed aseptically and streaked out on each of two blood agar plates one of which is incubated aerobically and the other anaerobically.

This procedure is repeated on alternate days for at least three weeks; all organisms thus isolated must be fully identified.

Various media are available for blood culture, e.g., saponin broth is of special value in isolating *Strept. viridans*, sodium taurocholate broth in cases of suspect enteric fever.

Basally these and all other media for blood culture contain *p*-aminobenzoic acid and penicillinase to nullify the activity of sulphonamides and penicillin should these have been used in treating the patient.

CLOT CULTURE

In the case of the enteric fevers clot culture yields results quantitatively superior to the blood culture technique described above.

For clot culture 5 ml. of venous blood are collected aseptically in a screw-capped bottle and allowed to clot; the separated serum is then aseptically removed and 15 ml. of 0·5 per cent. bile-salt broth containing 100 units/ml. of streptokinase are added to the clot. Rapid lysis of the clot occurs and the mixture is then examined as for blood culture.

The advantages of clot culture are that the clinician does not require a supply of special blood culture bottles, isolation of salmonellae is superior to normal blood culture techniques and a Widal test can be performed on the serum, thus giving a basal titre for the particular patient against which subsequent antibody determinations may be judged.

CHAPTER 25

CEREBROSPINAL FLUID

SPECIMEN COLLECTION

FLUID is withdrawn by lumbar puncture with full aseptic precautions. If the specimen cannot be delivered to the laboratory immediately, the container should be kept at 37°C; in any case, laboratory procedures should be undertaken within 2 to 4 hr of the specimen being obtained.

LABORATORY PROCEDURE

In acute pyogenic infections the fluid is usually turbid in appearance but can vary from a clear to a grossly purulent state—a clear fluid is often seen in cases of aseptic meningitis.

One ml. of uncentrifuged CSF should be removed aseptically and added to a tube containing 1 ml. of 0·2 per cent. glucose broth. This favours the growth of meningococci and pneumococci without disturbing the growth of *H. influenzae* or other, rarer, causal organisms. After incubation at 37°C for 18 to 24 hr, subculture is made to blood agar plates which are incubated similarly to the procedure outlined in the diagram.

A cell count should be performed and then the remaining fluid is centrifuged at 3000 r.p.m. for 5 min. and 2 ml. of supernatant fluid transferred to a sterile container and submitted for biochemical examination, i.e., the estimation of the glucose, protein and chloride content. The deposit is seeded on to two blood agar plates one of which is incubated aerobically and the other in an atmosphere of air plus 5 per cent. CO_2. A Gram film is made from the deposit and the remaining fluid incubated overnight.

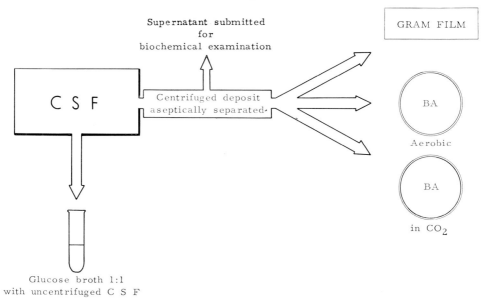

N. meningitidis appear as intracellular Gram-negative diplococci which may vary in number; pneumococci and *H. influenzae* also occur in clinically identical cases. Staphylococci and β-haemolytic streptococci are encountered in occasional cases and indeed almost all human pathogens have been isolated from CSF. The identification of the three commonly occurring organisms mentioned above must be fulfilled by cultural and biochemical methods and tested for sensitivity to antibiotic agents before a final report is issued but a tentative report can be made on the basis of the Gram-stained preparation.

Specimens from patients with suspected tuberculous meningitis are handled as indicated on page 160.

CHAPTER 26

URINE

Specimen Collection

CATHETERISATION is to be avoided whenever possible since it carries a definite risk of *causing* infection, either through faulty aseptic technique or by contamination with bacteria residing in the urethra. Certainly, catheterisation is never indicated solely to provide a specimen of urine for bacteriological testing; properly collected mid-stream specimens are entirely suitable in both sexes.

In males the prepuce should be retracted and the glans washed thoroughly; the first flow of urine flushes the urethra and reduces the resident bacterial flora. The mid-stream flow is collected in a sterile 2 oz. screw-capped bottle.

In females the vulva is washed with soap and water before the mid-stream specimen is collected in a sterile wide-mouthed container, e.g., a 1-lb. honey-pot.

Since many bacteria flourish in urine it is essential that the specimen be examined within 3 hr of collection; if this is not possible the specimen should be refrigerated immediately after collection and kept thus until laboratory examination can be undertaken.

Laboratory Procedure

Bacterial count. Dilutions of 1 in 100 and 1 in 1000 are made from the uncentrifuged specimen and incorporated in pour plates or spread over the surface of nutrient agar plates; after overnight incubation the resultant colonies are counted and the number of bacteria per ml. of specimen estimated. Counts of 10^3 organisms or less per ml. are indicative of contamination whereas counts of 10^5 organisms or more per ml. are associated with infection in the urinary tract. Specimens yielding 10^4 organisms per ml. may be difficult to interpret but where several bacterial species are present, such counts are most likely to mirror contamination and not infection; when counts in this region are made, further specimens should be examined to assist evaluation in a particular case.

Cultivation, etc. 5 to 10 ml. of the specimen are centrifuged at 3000 r.p.m. for 5 min. and the deposit resuspended in a suitable volume of supernatant. A wet film is made and examined microscopically for the presence of bacteria, pus cells and red blood cells; if this examination shows any bacteria a primary sensitivity plate is seeded from the deposit and discs of sulphonamide, streptomycin, chloramphenicol, tetracycline, etc., are applied. Finally, a blood agar and a MacConkey plate are inoculated with the deposit and incubated overnight. Following species identification, subculture sensitivity tests may, if necessary, be performed from the isolated colonies on the diagnostic media.

In the majority of cases of urinary tract infection only a single species is involved and the most commonly occurring are *Esch. coli, Pr. vulgaris, Pr. mirabilis* and *Ps. pyocyanea;* where two or more species are isolated then further freshly voided specimens should be examined. If the same species are repeatedly found together only then can it be assumed that both are involved in the infection.

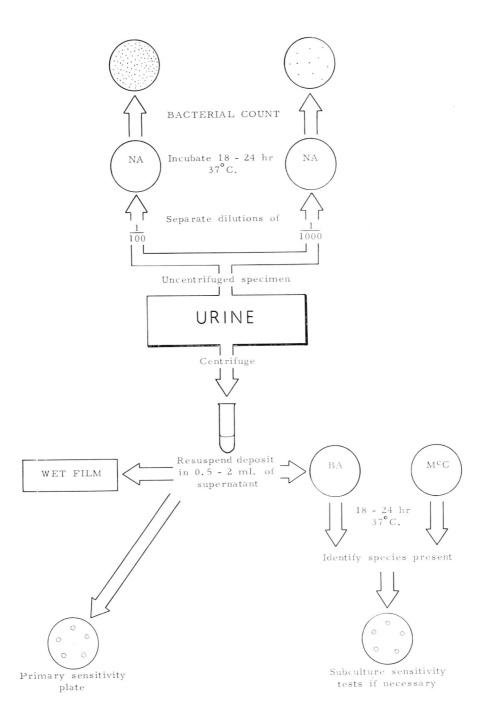

BACTERIAL COUNT

Incubate 18 - 24 hr
37°C.

NA

NA

Separate dilutions of

$\frac{1}{100}$

$\frac{1}{1000}$

Uncentrifuged specimen

URINE

Centrifuge

WET FILM

Resuspend deposit
in 0.5 - 2 ml. of
supernatant

BA

McC

18 - 24 hr
37°C.

Identify species present

Primary sensitivity
plate

Subculture sensitivity
tests if necessary

L

CHAPTER 27

THROAT SWABS

Specimen Collection

The throat should be examined with adequate illumination and a spatula must be employed so that the affected area can be swabbed accurately and without contamination from the buccal cavity; swabs should not be taken within 6 hr of gargling or of the administration of antimicrobial agents. If more than 12 hr delay is likely before the swab reaches the laboratory then serum-coated swabs are advantageous in ensuring the survival of *Strept. pyogenes*.

Laboratory Procedure

Strept. pyogenes, Vincent's organisms and *C. diphtheriae* must be sought in all throat swabs.

Strept. pyogenes. Crystal violet blood agar (CVBA) plates are inoculated; the incorporation of a 1 in 500,000 concentration of crystal violet reduces the growth of commensal organisms. A bacitracin-impregnated disc is placed in the well-inoculum of each plate; strains belonging to groups A, C & G are inhibited whilst those belonging to other groups grow freely in the immediate vicinity of the bacitracin disc. One of the plates is incubated anaerobically since this favours the growth of β-haemolytic streptococci in comparison with the rest of the throat flora; colonies of *Strept. pyogenes* are larger and show wider zones of β-haemolysis under anaerobic conditions.

C. diphtheriae. A Loeffler slope and a plate of tellurite medium are inoculated; growth from the Loeffler slope is rapid and films can be made after 12 to 18 hr incubation and stained by Gram's and Albert's methods. Volutin granules develop rapidly on this medium and the detection of Gram-positive bacilli which, with Albert's stain, show volutin granules demands subinoculation to blood agar and tellurite media so that colonies can be obtained in a pure state for further identification. Colonies on tellurite media may not be recognisable for 48 hr; if any colonies are regarded as those of *C. diphtheriae* they should be subcultured after their morphology has been studied. All suspect colonies must be subjected to fermentation tests and if these confirm their identity as *C. diphtheriae*, the final proof of their virulence is by demonstrating toxigenicity.

Vincent's organisms. A film is made from the throat swab and stained with dilute carbol fuchsin (1 in 10) for 10 min. The presence of *large numbers* of *Borr. vincentii* and fusiform bacilli is indicative of Vincent's infection.

With the above exception, stained films made directly from throat swabs are of no value in the diagnosis of throat infections since commensal cocci and bacilli will be found which morphologically are indistinguishable from *Strept. pyogenes* and *C. diphtheriae*. The swab should be used to inoculate the Loeffler slope, CVBA plates and the tellurite medium in that order so that the relatively inhibitory substances in the plate media are not carried over to the Loeffler slope. The film for Vincent's organisms is made last of all.

EXAMINATION OF THROAT SWABS

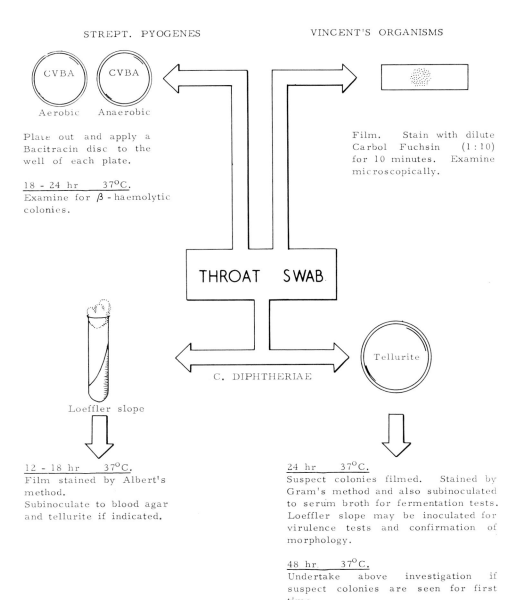

STREPT. PYOGENES

VINCENT'S ORGANISMS

CVBA   CVBA

Aerobic   Anaerobic

Plate out and apply a
Bacitracin disc to the
well of each plate.

18 - 24 hr   37°C.
Examine for β - haemolytic
colonies.

Film.   Stain with dilute
Carbol Fuchsin   (1:10)
for 10 minutes. Examine
microscopically.

THROAT   SWAB.

Loeffler slope

Tellurite

C. DIPHTHERIAE

12 - 18 hr   37°C.
Film stained by Albert's
method.
Subinoculate to blood agar
and tellurite if indicated.

24 hr   37°C.
Suspect colonies filmed. Stained by
Gram's method and also subinoculated
to serum broth for fermentation tests.
Loeffler slope may be inoculated for
virulence tests and confirmation of
morphology.

48 hr.   37°C.
Undertake   above   investigation   if
suspect colonies are seen for first
time

CHAPTER 28

SPUTUM

SPECIMEN COLLECTION

ONE of the most frequent frustrations in diagnostic laboratories is to receive a specimen of 'sputum' which is entirely salivary in origin. It is essential that the specimen should be coughed up and not merely result from expectoration of saliva or hawked from the post-nasal space. Many of the pathogenic species involved in chest infections are feebly viable outside the host's tissues so that the specimen should be delivered to the laboratory rapidly.

LABORATORY PROCEDURE

Since all respiratory tract pathogens, excepting tubercle bacilli, can be isolated (usually in small numbers) from the healthy upper respiratory tract and since by direct sampling of the raw specimen in different parts, different organisms may be cultured it is essential that the specimen be homogenised before plating; thus a clearer picture of the relative proportions of different bacteria can be obtained.

An equal volume of 1 per cent. buffered pancreatin is added to the sputum and the container thoroughly shaken; it is placed in a 37°C water bath and shaken every 15 min. for 1 hr. Alternatively, the specimen may be shaken up with sterile glass beads. Thereafter a Gram-stained film is made and examined microscopically for cellular exudate and bacteria.

Laboratory practice varies in regard to the media routinely inoculated with sputum specimens; the minimal requirement is cultivation on a blood agar plate incubated aerobically. In the well-inoculum are placed an 'optochin' disc for rapid differentiation of *Strept. viridans* and pneumococci and a penicillin disc which will inhibit the growth of many other organisms whilst allowing *H. influenzae* to flourish. A primary sensitivity plate should be inoculated in parallel with the diagnostic medium. Preferably a second blood agar plate should be inoculated and incubated in an atmosphere with added CO_2 (5 per cent.) since occasional strains of pneumococci and *H. influenzae* will not grow on primary isolation except in such an atmosphere.

If the sputum is from a suspect case of tuberculosis it will be processed as indicated on page 160; in such instances at least six consecutive daily specimens should be examined.

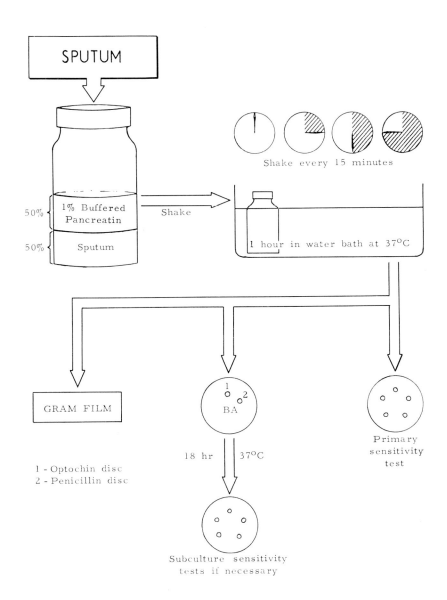

SPUTUM

Shake every 15 minutes

50%   1% Buffered Pancreatin   Shake

50%   Sputum

1 hour in water bath at 37°C

GRAM FILM

1 - Optochin disc
2 - Penicillin disc

1
2
BA

18 hr   37°C

Primary
sensitivity
test

Subculture sensitivity
tests if necessary

CHAPTER 29
FAECES

SPECIMEN COLLECTION

WHENEVER possible the specimen should be examined within a few hours of being passed; if more than 18 to 24 hr delay is likely before reaching the laboratory, the faeces should be added to an equal volume of buffered glycerol-saline solution and thoroughly mixed. Rectal swabs may be used in place of faeces in the investigation of suspect cases of bacillary dysentery but these must be rectal and not merely anal swabs and must be plated immediately if reliable results are to be obtained. Shigellae (particularly *Sh. sonnei*) are the commonest cause of bowel infections in Britain and all faecal specimens must be examined for these; the same media also serve the isolation of salmonellae. Laboratories vary in their attitude towards searching for the less common enteropathogenic *Esch. coli* types; some include this only when requested, others include such as part of their routine examination of all faecal specimens.

LABORATORY PROCEDURE

Any specimen which is at all formed must be thoroughly emulsified in sterile physiological saline before being plated on to diagnostic media.

Shigellae and Salmonellae. The specimen is plated out on selective media (DCA and MacConkey's) and also inoculated into tubes of selenite and tetrathionate enrichment broths. After overnight incubation these broth cultures are subinoculated to a fresh DCA plate. On either of the selective media, lactose non-fermenting (pale) colonies must be regarded as shigellae or salmonellae until proven otherwise; several such colonies from each plate should be submitted to biochemical tests. The inclusion of a tube of urea allows rapid elimination of urease producers which are not intestinal pathogens. The final type identification of a member of either pathogenic genus is by serological methods.

Enteropathogenic Esch. coli. In addition to inoculating a MacConkey and DCA plate (strains grow poorly if at all on the latter medium), a blood agar plate should be inoculated lightly since occasional strains grow only on this medium. There are no enrichment media for *Esch. coli*. On MacConkey's or BA media colonies of enteropathogenic strains are identical with all other types of *Esch. coli.*, nor is there any method of biochemical differentiation; therefore it is necessary to test, by slide-agglutination, at least 10 colonies from each plate with a polyvalent antiserum prepared against the six enteropathogenic types. If agglutination is noted then the reacting colony is further tested against each of the individual type-specific antisera.

It should be appreciated that the macroscopic characters of faeces in cases of Sonne dysentery are only rarely of the classical type, i.e., a mixture of mucus and blood; usually the stool is unaltered apart from its consistency. Whilst wet film preparations may reveal the presence of pus cells and/or red blood cells, Gram-stained preparations of faeces are of no assistance in diagnosis on account of the morphological similarity of pathogens and the commensal coliform flora.

154

Serologic investigation of at least 10 colonies must be undertaken: Colonies of enteropathogenic strains do not differ in appearance among themselves or from other coli strains; nor is there a significant biochemical differentiation.

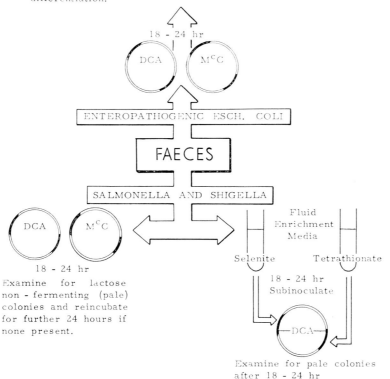

18 - 24 hr

DCA   M^cC

ENTEROPATHOGENIC ESCH. COLI

FAECES

SALMONELLA AND SHIGELLA

DCA   M^cC

18 - 24 hr

Examine for lactose non - fermenting (pale) colonies and reincubate for further 24 hours if none present.

Fluid Enrichment Media

Selenite   Tetrathionate

18 - 24 hr Subinoculate

DCA

Examine for pale colonies after 18 - 24 hr

Pale colonies are further investigated by inoculating into a differential row of sugar media.

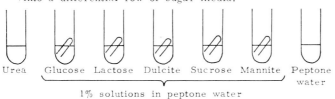

Urea   Glucose Lactose Dulcite Sucrose Mannite   Peptone water

1% solutions in peptone water

When a lactose non-fermenting bacillus with the general characteristics of either genus is isolated, its precise identification is obtained by serologic techniques.

CHAPTER 30

PUS AND WOUND EXUDATES

Specimen Collection

There are at least two reasons why a volume of pus or part of an excised wound is preferred for examination rather than swab samples of such material; firstly, the swab specimen will dry out to a greater or lesser extent and some pathogens, e.g., *Strept. pyogenes*, may not survive (serum-coated swabs reduce this hazard). Furthermore the number of media employed in examining such specimens implies a reducing inoculum over several plates and tubes of culture media. If swabbing is the only practical method of sampling then at least three swabs should be submitted; one of these being used to prepare films for staining and the others for inoculating media. If the lesion is deep-seated and draining through a sinus then the entire inner dressing should be sent to the laboratory in a sterile, wide-mouthed jar.

Laboratory Procedure

1. **Macroscopic examination** should always be made and if actinomycosis is a possible diagnosis, a particular search should be made for granules which may in early cases be atypical in appearance, i.e., white or even semi-transparent. When part of an excised wound or other tissue is submitted it must first be ground in a sterile pestle and mortar or in a tissue-grinding tube.

2. **Microscopic examination.** Gram-stained films must always be made and if granules are noted in the specimen one should be removed and crushed between two slides before staining. In cases of suspected tuberculous infection a Ziehl-Neelsen stained film should be made directly from the material and part of the specimen subjected to examination as outlined on page 160.

3. **Cultural methods** (*a*) **Pyogenic organisms.** Certain bacteria are very frequently found in pus-producing lesions, e.g., *Staph. pyogenes* and *Strept. pyogenes* and are thus termed pyogenic but it must be noted that many other bacteria are found on occasion in such situations, usually in association with an accepted pyogenic species. Cultures should be made on two blood agar plates, one incubated aerobically and the other anaerobically and, in addition, a plate of MacConkey's medium should be inoculated. Additional cultures may be made on other media depending on the microscopic findings and/or the clinical history of the patient; in any case a primary sensitivity plate should be set up.

(*b*) **Anaerobic organisms.** The most important species are members of the genus *Clostridium*; since such organisms may occur in wounds solely as contaminants without producing infection the diagnosis of tetanus or gas gangrene must be made on clinical grounds and treatment instituted without awaiting bacteriological confirmation.

156

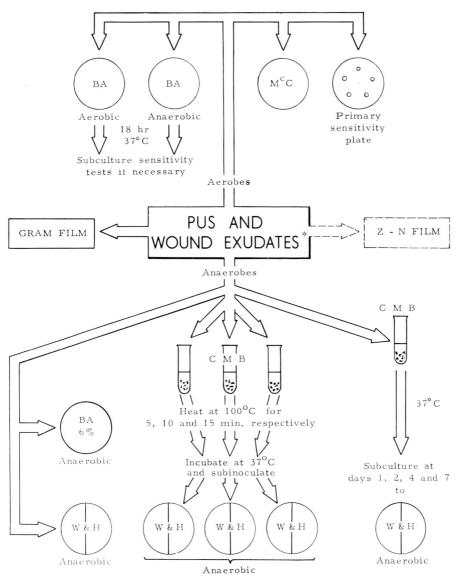

BA — Aerobic
BA — Anaerobic

18 hr 37°C

Subculture sensitivity tests if necessary

M^cC

Primary sensitivity plate

Aerobes

GRAM FILM

PUS AND WOUND EXUDATES*

Z - N FILM

Anaerobes

C M B

C M B

37°C

Heat at 100°C for 5, 10 and 15 min. respectively

BA 6%

Anaerobic

Incubate at 37°C and subinoculate

Subculture at days 1, 2, 4 and 7 to

W & H — Anaerobic

W & H — W & H — W & H

Anaerobic

W & H — Anaerobic

* Whenever possible, liquid pus (removed by syringe) and excised tissue from infected wounds should be submitted in preference to swabs.

157

A plate of blood agar (agar content increased to 6 per cent. to reduce the spreading tendency of *Clostridia*) and a plate of half-antitoxin Willis and Hobbs' (W & H) medium are inoculated directly from the specimen. The half-antitoxin W & H plate, using a mixture of *Cl. welchii* type A and *Cl. oedematiens* type A antitoxic sera smeared over one half of the plate, is specially valuable in identifying *Cl. welchii* and *Cl. oedematiens*.

Three tubes of cooked-meat broth (CMB) are inoculated from the specimen and heated at 100°C for 5, 10 and 15 min. respectively before being incubated at 37°C and subsequently subcultured at 24 and 48 hr to half-antitoxin W & H plates. Heating of these tubes prior to incubation kills non-sporing organisms and facilitates particularly the isolation of *Cl. oedematiens* and *Cl. septicum*. Finally another tube of CMB is inoculated from the specimen and incubated at 37°C, subcultures being made after 1, 2, 4 and 7 days; in this tube both aerobes and anaerobes flourish. The early subcultures are made for rapidly growing clostridia and the later ones for slower growing organisms; additionally 6 per cent. blood agar plates can be inoculated aerobically and anaerobically on these occasions. All cultures made on plated media must be incubated for at least 48 hr before being rejected as yielding no growth; plates should be examined after 24 hr incubation but this inspection should be brief since several clostridia, e.g., *Cl. oedematiens* and *Cl. tetani* die very rapidly on exposure to air.

CHAPTER 31

TUBERCULOUS INFECTION

SPECIMEN COLLECTION

TUBERCULOSIS can affect virtually every tissue in the body, thus the specimen submitted may vary widely. Whilst there should be no unnecessary delay in forwarding specimens for examination, the hardiness of tubercle bacilli allows them to survive very readily. Cerebrospinal fluid is prevented from clotting by adding sterile sodium citrate; serous fluids, e.g., pleural effusion are similarly treated. When urine is to be submitted for examination the specimen may be either a 24 hr collection or three pooled, consecutive, early-morning specimens; cleansing of the genitalia with soap and water will reduce the population of *M. smegmatis*.

LABORATORY PROCEDURE

Direct films from the specimen, stained by Z-N method are valuable only if the material contains very large numbers of bacilli.

Isolation of Mycobacterium tuberculosis

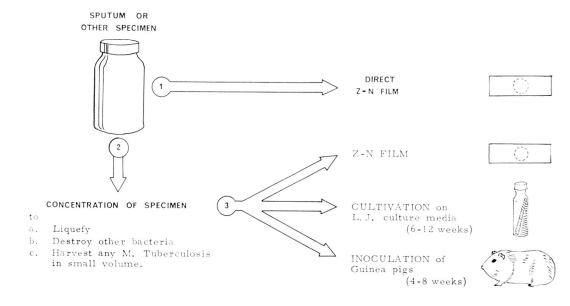

SPUTUM OR
OTHER SPECIMEN

1 → DIRECT Z-N FILM

2

CONCENTRATION OF SPECIMEN
to
a. Liquefy
b. Destroy other bacteria
c. Harvest any M. Tuberculosis in small volume.

3 → Z-N FILM

CULTIVATION on
L. J. culture media
(6-12 weeks)

INOCULATION of
Guinea pigs
(4-8 weeks)

All specimens which are not fluid and/or those which may contain other organisms must be subjected to concentration before further processing; numerous concentration techniques are available but all have the same aims, i.e., to liquefy the specimen thus releasing any tubercle bacilli, to destroy all other bacteria and, by centrifugation, to harvest any *M. tuberculosis* in a small volume. One concentration method is a modification of Petroff's technique: A volume of the specimen is mixed with an equal volume of 4 per cent. caustic soda and thoroughly shaken; the mixture is incubated at 37°C for 30 min. and the container shaken thoroughly every 5 min. After incubation and shaking the mixture is centrifuged for 30 min. at 3000 r.p.m. and the supernatant fluid is discarded; a drop of phenol red solution is added to the deposit which is then very carefully neutralised with 8 per cent. HCl.

The concentrate is used to make films to be stained by the Z-N method, and to inoculate slopes of Lowenstein-Jensen (L-J) media; cultivation must always be undertaken since it is much more reliable than microscopy and also the bacilli are then available for sensitivity testing. Guinea-pigs may also be inoculated with the concentrate; 0·5 ml. of the concentrate is injected into the left thigh muscles of an adult guinea-pig. Inoculation yields as high a proportion of positive results as does cultivation; with certain specimens, e.g., CSF and urine, inoculation yields a higher proportion of positive results than cultivation of the same material. Because of the risk of death from intercurrent infection it is usual to inoculate two guinea-pigs with material from each specimen.

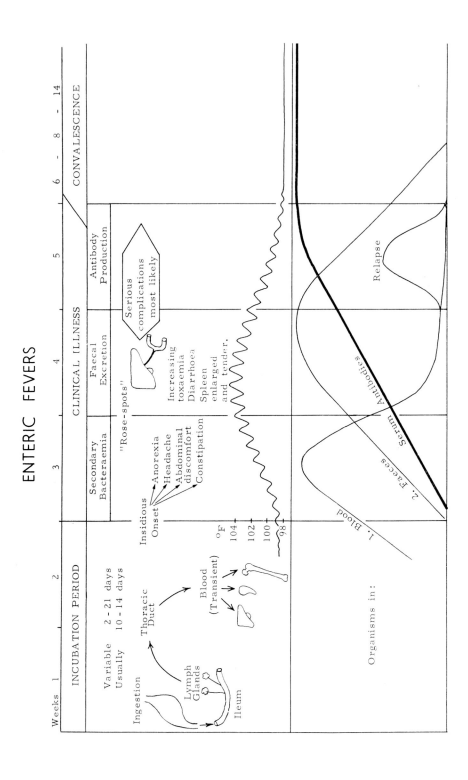

ENTERIC FEVERS

CHAPTER 32

THE ENTERIC FEVERS

THESE include typhoid and the paratyphoid fevers which in their essential clinical features are exactly alike; similarly their gross pathology is the same so that only by isolating and specifically identifying the causal organism can a distinction be made. *S. paratyphi B* is the commonest organism encountered in Britain and cases infected with *S. typhi* have usually acquired the infection overseas; *S. paratyphi A* and *S. paratyphi C* are rarely found in Britain.

The progression of clinical events illustrated is that of an untreated case; specific therapy, e.g., with chloramphenicol, rapidly eliminates bacteraemia, reduces the duration of pyrexia and alleviates rapidly the toxaemic symptoms.

The chart indicates the time of illness when isolation from the bloodstream and the faeces is most likely to succeed; microscopic examination of these specimens has no place in diagnosis.

BLOOD CULTURE

This is performed as in Chapter 24 and the clinician should obtain from the laboratory, blood culture bottles containing 0·5 per cent. sodium taurocholate broth which medium enhances the isolation of enteric fever bacilli. Blood culture is most likely to be positive during the first 10 days of clinical illness and during relapses; three specimens should be obtained at 6 to 12 hr intervals. It should be noted that within 3 hr of chloramphenicol therapy being instituted blood cultures are most unlikely to yield positive results.

Clot culture has been mentioned in Chapter 24 and the advantages of this method merit its wider use. *Bone marrow culture* is as reliable as blood culture and remains positive for 1 to 2 days after chloramphenicol therapy has begun but it is doubtful if such a procedure could be justified except in a minority of cases.

FAECES CULTURE

The causative organisms can be isolated from the stool throughout the course of the illness but are most readily and frequently isolated during the second and third weeks; it is important that repeated examination of the faeces is undertaken since this increases the proportion of positive isolations.

Bile culture. Bile, aspirated by means of a duodenal tube, may be cultured in an attempt to isolate the causative organisms; this method is perhaps best restricted to carrier detection.

URINE CULTURE

Bacilluria is intermittent in enteric fever cases and if attempts at isolation from urine are to be successful then daily examination of the morning urine for one week should be made. Even then, urine culture is not so frequently positive as stool culture; it is most likely to yield positive results after the second week of clinical illness. In the detection of carriers and determining whether carriage is intestinal and/or renal, care must be taken to ensure that stool specimens are not contaminated by urine since if the individual is solely a urinary carrier such an error might lead him to be regarded also as a faecal carrier.

Serological Evidence of Infection

The Widal agglutination technique is dealt with opposite; the reaction is usually positive by the 7th to 10th day and the titre reaches its maximum during the 4th week. The patient's serum should be tested against both O and H suspensions of each of the relevant serotypes since occasionally only O or H agglutinins are detectable, especially in the early stages of the infection.

Interpretation of Widal reaction. Depending on the country in which an individual lives and even on the community within any country, his serum may agglutinate the test suspensions in low dilution and such reactions are without diagnostic significance. In Britain the usual limits of such normal reactions are, for *S. typhi* and *S. paratyphi B*, H agglutination=1 in 30, O agglutination=1 in 50 and for the two other serotypes both O and H reactions =1 in 10. In a case of enteric fever such low titres will be commonplace in the first week of the clinical illness so that a second serum sample must be tested 7 to 10 days later and will show a significant increase in titre. Anyone inoculated with TAB vaccine also possesses specific agglutinins and these may complicate the interpretation of the Widal reaction; in such individuals a definitely rising titre for any one of the organisms might be regarded as diagnostically significant. However other non-enteric infections may cause an increase in titre which however falls rapidly on recovery unlike the maintenance of titre in enteric fever. The clinician should inform the bacteriologist of a history of TAB vaccination.

Quantitative agglutination test

Quantitative testing of serum for agglutinins towards bacteria is most frequently employed in suspect cases of the enteric fevers and contacts of such cases—the Widal test. Such tests can be applied in brucellosis and several other infections provided suitable standardised bacterial suspensions are available.

The patient's serum (P.S.) is initially diluted 1 in 15 with sterile physiological saline.

0·4 ml. of saline is pipetted into tubes 2–7 and then 0·4 ml. of 1 in 15 P.S. is pipetted into each of tubes 1 and 2, so that the serum dilution in tube 2 is now 1 in 30; serial dilution is now effected by thoroughly mixing the contents of tube 2 and then transferring 0·4 ml. into the unit volume of saline in tube 3 and so on up to and including tube 6, from which, after mixing the contents, 0·4 ml. is rejected. The dilutions now read from 1 in 15 through to 1 in 480 and the seventh tube, without serum, acts as a control in which only the bacillary suspension is added to ensure that the latter suspension is not auto-agglutinable. Finally, 0·4 ml. of the standard bacillary suspension is added to each tube in the series, using a fresh sterile pipette and working from tube 7 through to tube 1 so that if any serum dilution is carried over from one tube to another the effect on the dilution is less than if the carry-over were from the lower dilution to the next highest.

Finally, the contents of each tube are thoroughly mixed and then transferred by individual capillary pipettes to agglutination tubes which are then incubated in a water bath at 37°C for 4 hr.

Agglutination is indicated by clearing of the supernatant fluid to an extent depending on the amount of bacterial suspension which has deposited.

The titre of a particular serum is indicated by the tube containing the highest dilution of serum in which agglutination is noted by the naked eye.

QUANTITATIVE AGGLUTINATION TEST

UNIT VOLUME 0.4 MI

SALINE

REJECT
0.4 MI

1/15 1/30 1/60 1/120 1/240 1/480 C

1/30 1/60 1/120 1/240 1/480 1/960 C

B.S.

TRANSFER AND INCUBATE

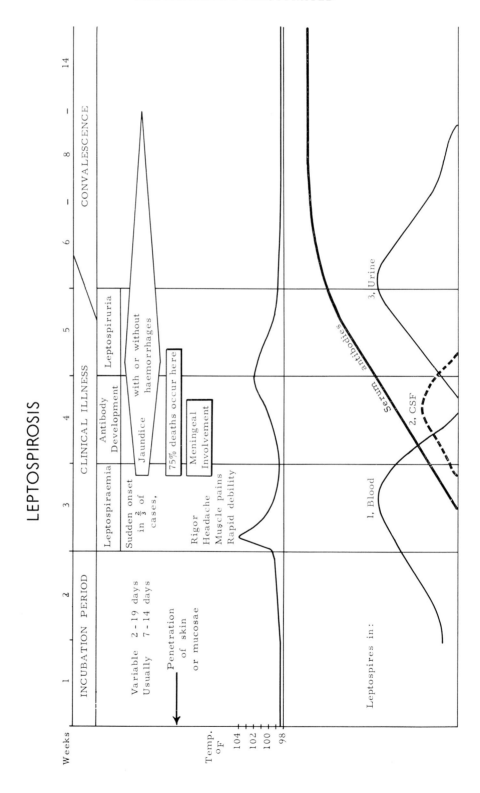

LEPTOSPIROSIS

CHAPTER 33

LEPTOSPIROSIS

THE accompanying chart outlines the clinical phenomena at various stages of Weil's disease but it must be noted that the disease varies in severity and also that inapparent infections may occur in individuals exposed occupationally to the risk of infection. Furthermore, leptospirosis caused by the thirty or more other members of the genus is usually much milder than Weil's disease and has a correspondingly lower case fatality rate. The chart also indicates the times at which leptospirae are most likely to be demonstrated in the bloodstream, cerebrospinal fluid and urine.

Because leptospirae are extremely sensitive to an acid environment it is essential to test the pH of urine specimens immediately after collection; if necessary the pH should be adjusted to a level not less than 7·5. With any of these specimens three laboratory procedures can be used.

MICROSCOPY

Either of stained films or by dark-ground illumination of the fresh specimen.

CULTIVATION

Various fluid media have been developed; after inoculation these are incubated at 37°C and samples are examined by dark-ground microscopy every 3 to 4 days for 4 weeks. As soon as growth is thus detected further incubation should be at 30°C.

ANIMAL INOCULATION

This is the most sensitive of the diagnostic methods. Young guinea-pigs or golden hamsters are inoculated intraperitoneally with the relevant specimen; 3 days after inoculation abdominal paracentesis is performed and the aspirated fluid examined by dark-ground microscopy. If actively motile leptospirae are seen then cardiac puncture should be performed and the heart-blood used to inoculate fluid media. If at this stage the peritoneal aspirate shows no leptospirae then paracentesis is repeated daily for 4 more days.

With strains of leptospirae which are fully virulent to the experimental animal, the latter becomes febrile 4 to 5 days after injection and shows roughening of its coat and loses its appetite; jaundice appears by the 7th to 8th day and the animal dies within another 12 to 48 hr.

Serological Evidence of Infection

Several techniques are available for the demonstration of specific antibodies in patients' sera. Agglutination tests with killed suspensions of the various serotypes are best suited for routine diagnostic purposes. It will be noted from the chart that antibodies are rarely detected until the second week of the clinical illness but a test should be carried out as early as possible so that a base-line titre for the patient can be determined and is available for comparison with later tests. As in all serological tests the most convincing evidence of recent infection is the demonstration of a rising titre when a second serum sample is tested 4 to 7 days after the first; however, a titre of 1 in 300 or more on first testing in a person with no occupational risk of previous subclinical infection can be regarded as diagnostic.

In people known to follow occupations which carry a risk of inapparent infection, titres of up to 1 in 100 can exist without clinical illness; in such cases repeated testing at 3 to 4 day intervals will show a steeply rising titre if they are infected whereas the titre will not rise in the absence of active leptospiral infection.

CHAPTER 34

MISCELLANEOUS INFECTIONS

For those infections not dealt with in the processing charts the following procedures are adopted.

WHOOPING COUGH

Nasopharyngeal or per-nasal swabs are plated out on Bordet-Gengou medium; in addition a 'cough plate' can be used—a plate of Bordet-Gengou medium is held 4 to 6 in. in front of the patient while he is coughing and is thus inoculated with the expelled droplets. Bacteriological confirmation by isolation of *Bord. pertussis* is of a high order in the first 2 to 4 weeks of illness, especially in the catarrhal stage preceding the onset of whooping, but thereafter the isolation rate diminishes rapidly. Most laboratories incorporate a small amount of penicillin in Bordet-Gengou plates (6 units/12 ml. of medium) thus reducing the growth of many other organisms and allowing any *Bord. pertussis* to be more easily recognised.

Complement-fixation and agglutination tests may be applied to patients' sera after the third week of illness but such tests are rarely carried out; their role is the retrospective diagnosis of the atypical or missed cases at a stage of disease when isolation of the causative organism is unlikely.

CONJUNCTIVITIS

Swabs are processed as for the isolation of aerobic organisms from pus (page 156) but additionally a heated blood agar plate should be inoculated and incubated in an atmosphere of 5 to 10 per cent. CO_2 to facilitate the isolation of *N. gonorrhoeae* and *H. influenzae*.

BRUCELLOSIS

During the febrile phases, blood culture should be performed; several such specimens should be obtained and the clinician should obtain Castaneda blood culture bottles. These incorporate a layer of liver-infusion agar along one of the narrow sides so that by tilting the bottle every 48 hr the blood-broth mixture flows gently over the agar and the bottle is then reincubated in the upright position; the agar surface is inspected daily for developing colonies. Incubation should be continued for at least **4 weeks** before cultures are discarded as negative. During the chronic stage of the disease, blood culture is valueless; biopsy material, e.g., lymph glands, may be cultured successfully. The diagnosis during the afebrile phase is essentially by agglutination tests with the patient's serum against standardised suspensions of *Brucellae*. Titres of 1 in 80 or more indicate present or past infection and a rising titre on testing a second specimen is conclusive of active disease.

VENEREAL INFECTIONS

GONORRHOEA. Smears may be prepared directly from the discharge but it is essential also to culture such material. In the male, the meatus is cleaned with sterile gauze soaked in saline and then the discharge is collected with a sterile bacteriologic loop and plated immediately; if this is impracticable then several capillary tubes should be used to collect material and after sealing with plasticine they are sent immediately to the laboratory. In the female, similar procedures are undertaken and additionally secretions from the cervix uteri should be examined; a vaginal speculum must be used in collecting cervical secretions and the material can be most readily collected in a Pasteur pipette. Diagnosis of chronic gonococcal infection and its complications may be undertaken by complement-fixation tests on the patient's serum; the results of such tests, whether positive or negative, must be interpreted very carefully since false reactions are not uncommon.

SYPHILIS. In the primary stage, serous exudate should be collected from the chancre after thorough cleansing of the area to reduce the population of other organisms; the exudate is collected in capillary tubes and delivered immediately to the laboratory for examination under the dark ground microscope. Refrigeration of such specimens is to be avoided since it immobilises *Tr. pallidum* and makes microscopic recognition very difficult. Similar procedures can be taken with exudate from skin eruptions and mucous patches in the secondary stage. At any time after 2 to 3 weeks from the onset of infection, diagnosis is best undertaken by testing the patient's serum for antibodies; several methods and techniques are available and details of these should be sought in a standard text. The interpretation of serological tests requires close collaboration between venereologist and bacteriologist.

The possibility of both infections being acquired simultaneously must be borne in mind and since the incubation period of gonorrhoea is brief (3 to 9 days) in comparison with that of syphilis (3 to 6 weeks) then antibiotic treatment aimed at curing the gonococcal infection may mask the syphilitic infection which might proceed to the secondary stage without the development of a primary lesion.

Therefore all cases of gonorrhoea should be submitted to serological surveillance to ensure that they did not also acquire syphilis.

CHAPTER 35

ANTIMICROBIAL SENSITIVITY TESTS

THERE are few infective diseases caused by a single bacterial species always sensitive to a particular antibiotic or chemotherapeutic agent. Indeed with the exception of *Strept. pyogenes* and *Pneumococcus* which are constantly sensitive to penicillin, no species can be assumed to be sensitive to a particular drug.

The need for laboratory control of antimicrobial therapy is widely recognised but perhaps in no other area of medicine does the practice fall so short of the preaching; in many areas clinicians, in hospital or general practice, who seek laboratory confirmation of their diagnosis let alone guidance in therapy, are in the minority and one is left with the opinion that many clinicians either never see patients suffering from infections or that they 'treat' such patients by guesswork. Such crystal-ball therapy is an abuse of antimicrobial drugs, nationally uneconomic, and is not guaranteed to succeed.

Some of this abuse of antibiotics can be laid on the laboratory which delays unnecessarily in reporting useful information; provided the clinician is aware of the limitations of rapid laboratory techniques he should appreciate that he can obtain preliminary guidance of therapy at the same time as he receives a statement of the organisms isolated from the specimen—usually within 18 to 24 hr of the material reaching the laboratory.

Before specifying available techniques there fall to be enumerated several basic factors which may influence the results and which must be standardised.

Incubation time. If this is prolonged beyond 18 hr certain unstable agents, e.g., chlortetracycline will appear less efficient.

Medium. In general the constituents of the medium have little effect on antimicrobial agents but some, novobiocin and to a lesser extent the polymyxins, have less activity in the presence of serum; in the case of sulphonamides, media for *in vitro* tests must be free of substances antagonistic to these drugs.

The pH of the medium must also be standardised to ensure reproducible results, e.g., the activity of streptomycin and neomycin is greatly enhanced in an alkaline medium as are erythromycin and other macrolide antibiotics, whereas chlortetracycline is more stable and hence more active in an acid environment.

Inoculum. Without specifying the number of organisms used for *in vitro* tests of sensitivity the results, expressed as x units or micrograms required to inhibit the organism, are meaningless.

DISC DIFFUSION TECHNIQUE

Diffusion techniques are less precise than tube dilution methods but are more speedily performed and allow the simultaneous testing of several drugs; furthermore, diffusion techniques can be employed in direct testing of the pathological material so that some indication of sensitivity can be given simultaneously with identification of the causative organism.

Discs, 6·25 mm. in diameter are punched from Whatman No. 1 filter paper and in batches of 100 in screw-capped bottles are sterilised in a hot air oven at 150°C for 1 hr. The desired antibiotic solutions are prepared in sterile distilled water and 1 ml. added to each bottle; since 100 such discs absorb this unit volume each disc will contain approximately 0·01 ml. There is no general agreement as to the amount of a particular antibiotic contained in such discs but the following amounts per disc are useful in routine tests: penicillin 1 unit; streptomycin 10 μg; chloramphenicol 50 μg; tetracycline 10 μg; erythromycin 10 μg.

In all diffusion tests of activity the depth of medium must be standardised by using a constant volume in a Petri dish of constant diameter and with a flat surface; such plates must be poured on a horizontal surface. The inoculum may be spread uniformly over the surface with a loop or a broth culture or suspension may be flooded over the whole surface with a Pasteur pipette; thereafter, the plate is tilted and excess inoculum removed with the pipette. The inverted plates are left to dry on the bench for 1 hr before discs are placed on the surface with sterile forceps. Ideally a period of 3 hr prediffusion should be allowed before the plates are incubated. Diffusion tests may be performed using reservoirs other than discs for the antibiotics, e.g., a ditch may be cut from the medium and replaced with the antimicrobial agent mixed with agar, or cylinders (porcelain or steel) may be placed on the medium and the required agent pipetted into these.

Standard graphs have been prepared for each antibiotic which show the zone diameters of inhibition given by various concentrations of the antibiotic against a standard organism and these graphs allow the results of disc diffusion tests to be interpreted quantitatively.

The graph for penicillin is shown below and by measuring the zone of inhibition of an organism under test with a disc of known activity one can calculate the sensitivity of that organism to penicillin in comparison with that of the standard organism.

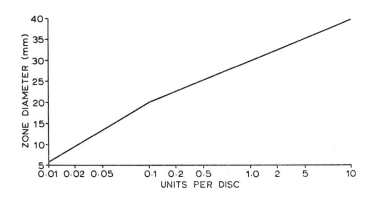

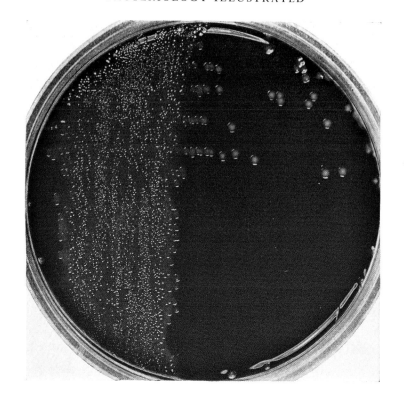

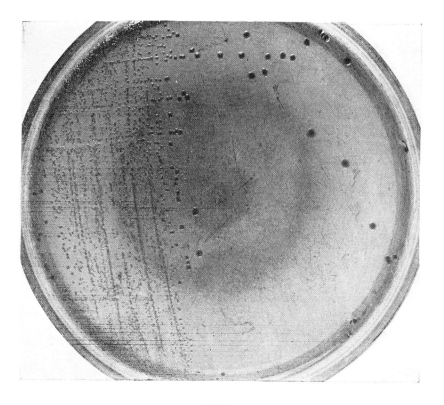

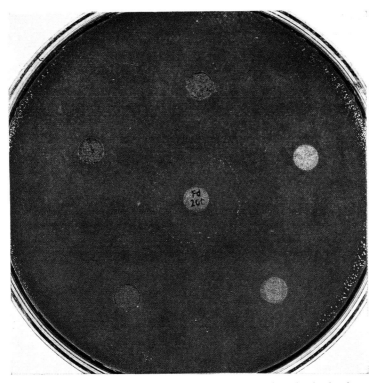

These three plates were inoculated from a centrifuged deposit of urine from a case of chronic pyelonephritis. On the left are shown the blood agar and MacConkey plates after overnight incubation at 37°C; these yielded a pure culture of *Esch. coli*.

The colonies on the blood agar plate are similar to those of many other enterobacteria; the growth on MacConkey's medium shows the characteristic, pink, lactose-fermenting colonies of *Esch. coli*.

Because of the presence of large numbers of bacilli and leucocytes in the wet-film preparation of the urine a primary sensitivity plate was seeded with the centrifuged deposit and, after the application of antibiotic discs, this plate was incubated in parallel with the diagnostic plates. Examination reveals that the strain is sensitive to chloramphenicol (green), streptomycin (white), chlortetracycline (yellow) and furadantin (centre disc) but resistant to sulphonamide and to penicillin (pink). Thus the clinician can be given a guide to therapy at the same time (within 24 hr of the laboratory receiving the specimen) as an identification of the causative organism.

175

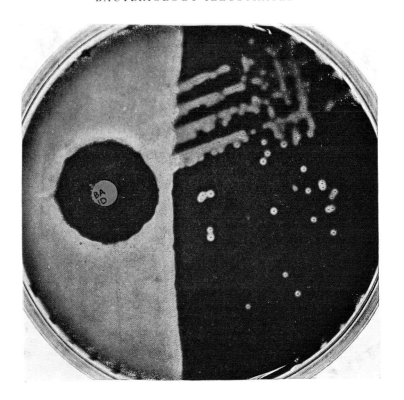

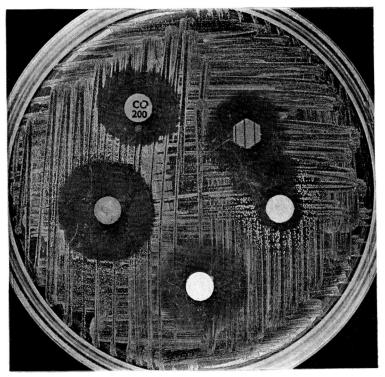

Subculture sensitivity test. Nutrient agar plate seeded with *Sh. sonnei* and, after the application of antibiotic discs, incubated overnight at 37°C. The strain is sensitive to colomycin (blue), chloramphenicol (green) and streptomycin (white); it is also sensitive to paromomycin (orange) and chlortetracycline (yellow) but these agents are also *synergistic* as indicated by the obliteration of growth in the area between their respective zones.

On the opposite page the upper illustration shows the use to which an antibiotic may be put for diagnostic purposes; this crystal-violet blood agar plate was inoculated from a throat swab. A bacitracin disc was placed in the well inoculum and after 18 hr incubation at 37°C a pure growth of β-haemolytic streptococci was obtained; sensitivity to bacitracin indicates that the strain belongs to Lancefield group A or more rarely to groups C and G, and thus facilitates rapid recognition of the groups most commonly pathogenic to man.

The lower illustration shows a subculture sensitivity test of the group-A strain, and penicillin (pink), chloramphenicol (green), chlortetracycline (yellow) and streptomycin (white) are all active against the organism. Sensitivity tests are not usually performed against *Strept. pyogenes* since all strains are eminently sensitive to, and readily eradicated by, penicillin.

177

Tube Dilution Technique

Tube dilution testing requires the preparation of two-fold dilutions of the antibiotic in a suitable fluid medium; a constant volume of the organism is added to each tube and a control tube containing no antibiotic is included in each test. After incubation at 37°C for 18 to 24 hr the tubes are examined for turbidity and that with the highest dilution showing no visible turbidity is the *bacteriostatic* concentration. The *bactericidal* concentration can be determined only by subculturing from those tubes without visible growth and transfer may be to agar plates or into broth devoid of antibiotic; the highest dilution yielding no growth on subculture is the bactericidal concentration. The tube test is too expensive of time, labour and materials for general purposes and it is normally reserved for testing the sensitivity of slow-growing organisms, particularly *M. tuberculosis*.

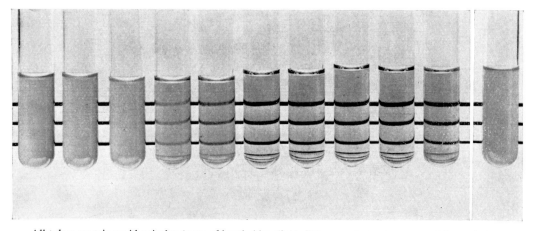

All tubes contain an identical volume of broth (the slight differences in level are caused by variation in the internal diameters of the tubes). Doubling dilutions of an antibiotic were prepared in the series so that tube 1 contains the highest dilution and tube 10 the lowest dilution of antibiotic; tube 11 is a control tube of broth without antibiotic.

To each tube was added an identical volume of an overnight broth culture of *Shigella sonnei*; after 18 hr incubation at 37°C, macroscopic examination revealed that growth had occurred in tubes 1–5 and 11.

The bacteriostatic concentration of the antibiotic thus lies between the amounts present in tubes 5 and 6; to determine the *bactericidal* concentration, aliquot volumes from each of tubes 6–10 were seeded on to a nutrient agar plate. After 18 hr incubation, growth was noted on the sector seeded from tube 6 and scanty growth on that inoculated from tube 7.

The *bactericidal* concentration of the antibiotic lies between those concentrations present in tubes 7 and 8.

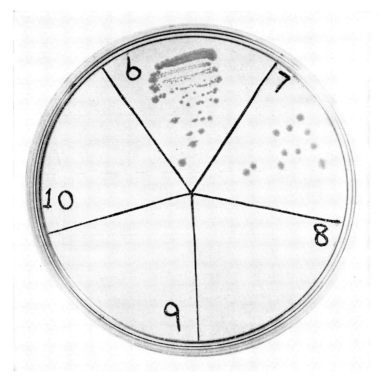

Determination of bactericidal concentration
This nutrient agar plate was seeded from tubes 6–10 in the dilution series and after 18 hr incubation, growth appeared only on these sectors inoculated from tubes 6 and 7.

ASSAY OF ANTIBIOTIC LEVELS IN BODY FLUIDS

There are three situations when it is necessary to establish the concentration of an antibiotic in human body fluids, firstly, *to confirm that adequate levels of antibiotic are being attained*, e.g., in the CSF of a patient with meningitis or more generally when an antibiotic is being administered orally to a patient whose intestinal tract may be deficient in its absorptive ability.

Secondly, *to ensure that blood levels do not exceed the limits* at which certain antibiotics, e.g., streptomycin, may cause nerve deafness and finally in *the investigation of a new antimicrobial agent*, assay of levels in body fluids allows the determination of its adsorption, distribution within the body and its rate and route of excretion.

Assay may be performed by either a diffusion technique or tube dilution method; if the material to be assayed for antibiotic content is likely to be contaminated with microorganisms these must be removed by filtration or some other method before tests are made by the tube dilution method. In this method doubling dilutions of the material are prepared in nutrient broth and each tube is then seeded with equal volumes of a standard organism of known sensitivity to the agent being assayed. A control set of tubes is set up containing not only

179

the antimicrobial agent under assay but containing also a fresh sample of the same body fluid free from the antibiotic so that any natural bacteriostatic activity of the particular body fluid occurs in the control as well as in the test series of tubes. Both sets of tubes are then incubated and by comparing the dilutions which give inhibition of the growth of the standard organism one can estimate the amount of antimicrobial agent in the patient's fluid.

Assay by the diffusion method is carried out using porcelain cylinders in place of discs; once the procelain cylinders have been sealed into plates seeded with a test organism of known sensitivity to the antibiotic being assayed they are filled with dilutions of the fluid under test.

After incubation the significance of the zone diameters of inhibition are read from a standard graph prepared for the particular antibiotic.

One advantage of the cylinder diffusion method is that contaminated material, even faecal suspensions, can be assayed without attempting to get rid of microorganisms.

These brief comments silhouette the microbiological methods of assaying the levels of antibiotics in body fluids. Many other methods, chemical as well as biological, are available and for details of these a suitable reference book (e.g., Kavanagh, 1963) should be consulted.

CHAPTER 36

PROPHYLACTIC IMMUNISATION

REFERENCE has been made to immunisation procedures in various chapters in the second section of the text and in this chapter further synoptic notes on such procedures are offered.

Immunity is the ability of the animal body to resist infection with micro-organisms and the harmful effects of their toxins.

INNATE IMMUNITY

Every animal, including man, possesses natural barriers that protect it against microbial attack; these comprise the mechanical and chemical barriers of the skin and the various mucous membranes. Additionally phagocytosis, either by the wandering phagocytes of the blood or the fixed phagocytes of the reticulo-endothelial system, is in the front line of the non-specific defence mechanisms.

Other natural but somewhat more specific defence mechanisms include the antagonism displayed by the commensal flora against potential invaders, e.g., *Lactobacillus acidophilus* (Döderlein's bacillus) lives commensally in the vagina and there creates a highly acid secretion by fermenting the glycogen of the vaginal epithelium; similarly lactobacilli which are commensal in the gut probably exert a protective effect since, when they are eliminated by the use of broad-spectrum antibiotics, severe and sometimes fatal infections occur. These are usually caused by strains of *Staph. pyogenes* which have become resistant to the broad-spectrum antibiotics or by *Candida albicans*.

Properdin, a high molecular weight protein which is present in normal serum, is another component of natural immunity but its activity is directed against Gram-negative bacteria; other basic proteins with an antibacterial action are produced from tissue and blood cells which have been damaged by infection, e.g., spermine, which is lethal to tubercle bacilli.

ACQUIRED IMMUNITY

In addition to the natural protection against infection noted above, an individual may acquire immunity of a highly specific type, i.e., protection against a particular bacterium, or its toxins.

Such immunity may be acquired naturally or artificially.

Natural acquired immunity may be obtained *actively*, i.e., when the individual's tissues produce antibodies specific for an organism which has infected them, and such active immunity is usually long-lasting. Natural immunity may also be acquired *passively* by transplacental donation of antibodies from the mother to the foetus; passive naturally acquired immunity is of short duration and rarely lasts more than 3 to 6 months since the antibodies are foreign to the infant's tissues and are eliminated.

Artificial acquired immunity can also be obtained actively or passively; active artificial immunity follows the administration of an antigen against which the individual's tissues produce a specific antibody and, as in active immunity acquired naturally, the protection thus obtained is long-lasting in comparison with the fleeting passive immunity obtained by the injection of antibodies which have been obtained from some other animal and rarely lasts more than 4 to 6 weeks.

Prophylactic immunisation, i.e., the acquisition of immunity to microbial infection by artificial methods is one of the most spectacular successes in modern medicine and has been responsible for the virtual elimination of mortality in several diseases.

The vast majority of the agents which we now use to induce immunity artificially have been introduced in the present century as a logical development of microbiological research, but active artificial immunisation had been practised, particularly against smallpox, for very many years. The earliest recorded account was of variolation which was introduced to Britain 250 years ago by Lady Montagu who was the wife of the British Ambassador in Turkey; variolation, i.e., the administration by scarification, of material from a smallpox vesicle into the skin of a healthy subject who then usually suffered and survived a modified form of the disease, had been common practice in Turkey for decades. However, the disease resulting from variolation was often severe and sometimes fatal and in addition variolated subjects frequently acted as sources of epidemic infection, thus variolation was made illegal in Britain in 1840. Perhaps the main reason for banning variolation was that Jennerian vaccination, introduced in 1798, was much more successful and certainly less dangerous.

The only other attempt at prophylactic immunisation which preceded the discovery of microorganisms was in the case of measles; here on a more restricted basis Francis Home reported in 1758 his method of implanting, by skin incision, threads soaked in material from measley children in an endeavour to produce a modified form of the illness which would result in protection from the natural form of the disease which at that time carried a high mortality rate.

Fortunately the introduction of the hypodermic syringe (1853) preceded the discovery of microorganisms and our ability to offer active and passive immunisation to microbial infection.

SAFETY FACTORS IN IMMUNISATION PROCEDURES

The syringe

Like most other discoveries the hypodermic syringe has its disadvantages and these are associated with the possible transmission of infection. Unless a syringe and needle are *sterile* then sepsis is a constant risk following injection, not only of immunising agents but of any fluid; an even more frightening risk is that of the transmission of the virus of serum hepatitis from person to person if a syringe and needle are used communally, e.g., at immunisation clinics. The lengthy incubation period, on average 80 days (range = 40 to 160 days), following the injection of the causal virus (Virus B) frequently prevents the correlation of the infection with a particular injection unless an outbreak occurs, e.g., when several patients develop serum hepatitis and all have a history of venepuncture at a particular clinic on a given day; serum hepatitis occurs when even minute amounts of blood, as little as 0·01 ml., are transmitted from a case or carrier to another individual, e.g., by a lancet used for finger pricking to obtain small volumes of blood for blood cell counts.

Perhaps it is not unnatural that the incidence of sepsis and serum hepatitis following on injection is not known since such iatrogenic episodes are rarely published.

However, both can be avoided if *a separate sterile syringe and needle* are used for each individual receiving an injection.

Other human errors

The production of immunising agents is rigorously controlled and tests of the safety and potency of each batch of an agent are performed before the agent is issued for use. However, human frailty has led to occasional disaster; in the early days of BCG production when, in Germany, the vaccine was produced in a laboratory where virulent, human-type tubercle bacilli were also being grown more than 200 babies were accidentally given virulent bacilli instead of BCG and almost one-third of the babies died of acute tuberculosis. Much more recently and when more stringent procedures have been instituted for the control of vaccine, catastrophes still occur, e.g., the distribution of 'killed' poliomyelitis vaccine which contained live virulent polioviruses.

UNAVOIDABLE COMPLICATIONS OF PROPHYLACTIC IMMUNISATION

When an infection is endemic and rampant in a community and then immunisation against the infection becomes available, the risk of the complications of immunisation is acceptable to that community; however, when the particular infection has been eradicated, primarily as a result of active artificial immunisation, a stage is reached when more people suffer side-effects, and death, as a result of preventive inoculations than the number who die from the actual disease. At this point, the risk of prophylactic immunisation becomes unacceptable so that fewer people may be protected artificially and the community then becomes more and more susceptible to the natural disease.

Only by a continuing programme of health education and stressing the need for immunisation can we hope to remain free from the return, in an endemic form, of many infectious diseases such as smallpox and diphtheria.

As examples of the complications of active immunisation we can consider smallpox vaccination, pertussis immunisation and provocation poliomyelitis.

Complications of smallpox vaccination. As always there is the risk of sepsis at the site of vaccination unless an aseptic technique is followed which minimises the possible introduction of pathogenic bacteria. The serious complications include post-vaccinial encephalitis and generalised vaccinia; these potentially lethal complications occur most often following primary vaccination and particularly if this is delayed until adolescence or adulthood. The incidence of these complications is lowest when primary vaccination is undertaken between the first and fourth years of life and fatality rates for such complications in children vaccinated at this age are extremely low; therefore in countries such as Britain where smallpox is no longer endemic, vaccination should be deferred until after the first birthday.

Complications of all kinds are less frequent when the vaccine is administered by the multiple pressure method which has the additional advantage over the older scarification method of giving a higher percentage of successful vaccinations.

Complications of pertussis immunisation. The incidence of febrile convulsions in children immunised against pertussis is estimated to be 0·12 per cent. and probably less than a quarter of these cases are associated with immunisation since only this proportion follows within 72 hr of an injection.

By comparison, in unimmunised children a history of convulsions can be expected in approximately 5 per cent. before the fifth birthday; thus the risk of post-immunisation febrile convulsions is not great and the complication is rarely fatal. However, children with a history of convulsions should not be given pertusis antigen.

A more serious but much less common complication of pertussis immunisation is encephalopathy; the size of the risk is not accurately known but it must be very small when one considers that millions of children have been immunised and the number of reported cases of encephalopathy is few. On the other hand the incidence of encephalopathy as a complication of the natural disease is 0·8 per cent. and this risk is obviously much greater than that following immunisation.

Provocation poliomyelitis. It is only 17 years since the possibility of poliomyelitis being provoked by immunisation procedures was first mooted; confirmation of this suspicion was rapidly forthcoming and it is estimated that the overall risk of provocation poliomyelitis following prophylactic injections is one case per 37,000 injections. The incidence of provocation poliomyelitis is highest with certain combined antigenic preparations and particularly those containing alum.

The risk can obviously be minimised by ensuring that a child receives oral poliomyelitis vaccine early in his immunisation programme.

EFFICACY OF PROPHYLACTIC IMMUNISATION

With certain agents the high degree of protection is easily seen but in general the results of laboratory tests in experimental animals must be confirmed by controlled field trials in the human population; in the past this ultimate proof has not always been undertaken and occasionally we have been misled by animal experiments.

As an example of the importance of field trials in evaluating the efficacy of vaccines we can note in summary fashion the history of immunisation against the enteric fevers.

TAB vaccines. Immunisation against typhoid fever was introduced in 1897 and in 1916 the vaccine also incorporated antigens to offer protection against infection with *S. paratyphi A* and *S. paratyphi B*. This TAB vaccine was heat-killed and phenol was used as the preservative—'phenolised vaccine'; it had never been subjected to properly controlled evaluation in man.

In 1934 a new type of TAB vaccine—'alcoholised vaccine'—was reported to have a much higher protective value for mice by virtue of its Vi or virulence antigen which stimulated the formation of Vi antibody; in 1941 it was reported that human beings who had received alcoholised vaccine frequently possessed Vi antibodies which were rarely produced in men receiving phenolised vaccine.

However, the relative efficacy of these two types of TAB vaccine was not determined until field trials were undertaken in Yugoslavia in 1954; as a result it was shown that the incidence of typhoid fever in a population receiving phenolised TAB vaccine was 6·1 per 10,000 as compared with an incidence of 14·1 per 10,000 individuals who had received the alcoholised vaccine. The incidence of typhoid fever in a control group who received a 'vaccine' comprising *Shigella flexneri* was 19·2 per 10,000 individuals.

Thus it became obvious that no matter how efficient the alcoholised vaccine is in protecting mice against infection it had a very low protective value for men.

Similar field trials have since taken place in other countries where the enteric fevers are endemic and the reasonable degree of protection offered by the phenolised vaccine has been confirmed; in these trials a recently developed acetone-killed TAB vaccine was used in place of the alcoholised vaccine and its protective power was shown to be even greater than that of the phenolised product.

Prevention of tuberculosis. Whilst it is accepted that several factors are involved in the elimination of tuberculous infection, e.g., early detection and treatment of cases, tracing of contacts, provision of dairy herds free from bovine infection and/or pasteurisation of milk supplies, etc., the role of preventive immunisation with BCG or Vole vaccines is important.

The evaluation of these vaccines in field trials in Britain has been perhaps the most exciting work in post-war endeavours in prophylactic immunisation. These trials, instituted by the Medical Research Council in 1949, have clearly shown that either vaccine gives an 80 per cent. reduction in the incidence of natural disease and in particular they protect against the more severe forms of infection, i.e., tuberculous meningitis and miliary tuberculosis which in the pre-antibiotic era were almost invariably fatal.

Diphtheria prophylaxis. As stated earlier the high degree of protection offered by certain agents is so dramatic that there is no need to mount controlled trials. The prevention of diphtheria by mass immunisation with toxoid preparations is probably the best example of such dramatic prophylactic influences.

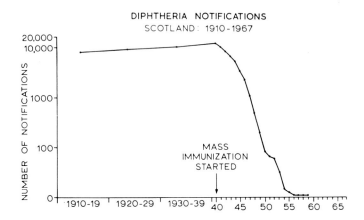

Apart from the almost incredible reduction in incidence as exemplified in the figure it should be noted that in Scotland for example no immunised person has died of diphtheria since 1948 and that only four confirmed cases have occurred in the last 10 years, all of whom had not been actively immunised.

Tetanus prophylaxis. Similar evidence for the high protective value of tetanus toxoid has been gathered over the last 50 years during wars; the low incidence of tetanus in civilian life makes it impracticable to undertake controlled field trials in most countries. Nevertheless, when one compares the incidence of tetanus in battle casualties in three periods, (1) before any form of immunisation was available, (2) when passive immunisation could be given after wounding and (3) in the Second World War when active immunisation was performed, then the spectacular drop in incidence can be noted.

	Cases of tetanus/1,000 casualties
Pre-immunisation	3–10*
Passive immunisation (after wounding)	1·5
Active immunisation (before wounding)	0·1

* Variation in incidence depended on the type of terrain on which fighting took place, i.e., highest on cultivated manured land.

PASSIVE ARTIFICIAL IMMUNISATION

The brief duration of passive artificial immunisation has already been noted and the short-term nature of such protection is dictated by the fact that the antibodies have almost invariably been derived from animals other than man and the human tissues recognise the complete foreignness of the animal globulins which are rapidly eliminated in a few weeks.

An additional disadvantage of passive immunisation with antisera derived from animals is the risk of hypersensitivity reactions to animal serum, e.g., immediate anaphylactic shock may occur in individuals who frequently have a personal or family history of allergy or serum sickness may occur as a delayed phenomenon some 1 to 2 weeks after administration of the antiserum. Such hypersensitivity reactions nowadays occur most commonly following the administration of tetanus antitoxin to an injured person who has not been actively immunised with tetanus toxoid and their occurrence is one of the main arguments for active immunisation against tetanus even although the disease is not common in civilian life.

IMMUNISATION SCHEDULES

The diseases against which active immunisation is presently available are listed in the table; it is indeed fortunate that for ecological reasons no one country requires to subject its citizens to protection against all of the diseases listed and these diseases against which people living in Great Britain should be protected are indicated by asterisks.

Diseases against which active immunisation is practicable

Cholera	Smallpox*
Diphtheria*	Tetanus*
Influenza	Tuberculosis*
Measles	Typhoid fever
Pertussis*	Typhus fever
Plague	Yellow fever
Poliomyelitis*	

Active immunisation can also be offered against certain other diseases but such protection is restricted to individuals at particular risk, e.g., those with an occupational risk of acquiring infection; these infections are listed below.

Anthrax	Mumps
Brucellosis	Q-fever
Leptospirosis	Tularaemia

In constructing a policy for immunisation in a particular country, priority must be given to preventing these infections which feature most frequently as causes of morbidity and mortality; for example, in Nigeria the incidence of smallpox and tuberculosis dictates that immunisation against these infections should be undertaken in the first four months of life and it is now known that

BCG and smallpox vaccines can be administered simultaneously but in different arms. This latter fact underlines the need to minimise the number of visits of children at immunisation clinics; in Africa where parents may bring their children many miles for immunisation it would be unrealistic to expect them to pay two visits when only one is necessary for protection against smallpox and tuberculosis.

Even in more sophisticated communities such as Great Britain the tedium of repeated visits to immunisation clinics has contributed to a fall-off in the proportion of the population who seek active immunisation and has stimulated the search for combined prophylactic agents as well as the creation of immunisation schedules which offer maximum protection with the minimum of inconvenience to mother and child.

The schedule outlined below is suggested for countries where public health services are satisfactory.

Immunisation Schedule

Age	Visit	Vaccine	Time interval
2–6 months	1	Triple	4–6 weeks
	2	Triple	4–6 weeks
	3	Triple	4–6 weeks
7–12 months	4	Polio (oral)	4–6 weeks
	5	Polio (oral)	4–6 weeks
	6	Polio (oral)	4–6 weeks
18–21 months	7	Triple	
2nd–4th year	8	Smallpox	
5 years	9	Dip./Tet.	
9 years	10	Dip./Tet./+Smallpox	
15 years	11	BCG	

The triple vaccine comprises pertussis, diphtheria and tetanus antigens and has priority in Britain because although pertussis is not frequently encountered in the first year of life more than half of the deaths from whooping cough occur before the first birthday.

Finally we must reiterate the need to pursue constantly intelligent methods of health education, so that mothers appreciate that only by continued artificial immunisation can we hope to prevent many communicable diseases from re-establishing themselves in Britain; this is particularly pertinent in the case of infections which were once endemic, e.g., diphtheria, but are now so rare that most young parents are aware of them only from family folk-lore.

SECTION IV

PROTOZOOLOGY

CHAPTER 37

FORTUNATELY protozoal infections are much less common in Britain than in certain other, mainly tropical, countries; this section offers some information about a few of the more important species of protozoa that are pathogenic to man.

Protozoa have several features in common with bacteria but differ from the latter so significantly in several respects that it is suggested that protozoa should be regarded as closer to mammals than to bacteria.

Like bacteria they are unicellular microorganisms but are usually *much larger* (2 to 100μ); whilst some protozoa reproduce by simple binary fission others undergo *schizogony* or multiple fission, i.e., as the cell grows the nucleus divides repeatedly so that the cell becomes multinucleate and only when growth ceases does cytoplasmic fission occur with the emergence of numerous uninucleate *merozoites*. Sexual reproduction occurs in some parasitic protozoa, e.g., malaria parasites, where zygotes are formed from the union of non-motile macrogametes with small, motile microgametes.

Protozoa also differ from bacteria in usually ingesting foodstuffs in particulate form and digestion proceeds within food vacuoles whereas much of the enzymatic activity of bacteria occurs at the cell surface; the organisation of the protozoal cell is mammalian in nature since within the cytoplasm there is a nucleus which contains DNA combined with protein to form chromosomes. The mitochondria of the protozoal cell resemble those of higher animals in structure and function and another similarity with mammalian cells is that protozoa can resist the action of antibiotics at concentrations which destroy bacteria.

Many protozoa have complex life cycles both in the human or animal host and in insect vectors; several species can form cysts, a stage in the life cycle when the organism is surrounded by a definite membrane. Cysts represent a resting phase in which the organism is much more resistant to adverse environments than are trophozoites.

ENTAMOEBA HISTOLYTICA

Several species of amoebae can inhabit the human gut but only *Ent. histolytica* is pathogenic to man; it can also infect other primates. Man is infected by ingestion of the cyst form either by direct contact with a carrier or case or indirectly from contaminated water supplies or from food contaminated by an infected food-handler. Insects can also transmit cysts from excreta to foodstuffs.

After ingestion by the host those cysts which escape the gastric juices pass into the small intestine where they excyst or hatch to form vegetative cells, i.e., trophozoites, which colonise the large intestine, particularly the caecum and pelvic colon.

Trophozoites may remain free-living in the intestinal lumen and in turn encyst and are excreted in the faeces, often in large numbers; alternatively they may invade the intestinal wall and cause ulceration. Classically amoebic dysentery runs a chronic course in comparison with bacillary dysentery and from the primary site in the intestine *Ent. histolytica* may spread by the lymphatics and bloodstream to the liver to give rise to amoebic abscesses; more rarely metastatic spread to other organs, e.g., lungs and brain, may occur. *Cysts are never found in infected tissues.*

The *trophozoite* measures 20 to 40μ in diameter and contains a single nucleus; this vegetative form is highly active, divides by binary fission and ingests red and white blood cells and tissue cells.

The *cyst form* of *Ent. histolytica* is less than half the diameter of the trophozoite and is spherical with a single smooth wall and is transluscent; cysts contain one, two or four but *never more than four nuclei.* Thick ovoid or rod-shaped structures known as chromatoid bodies or chromidial bars may also be seen.

Laboratory diagnosis. Although cultural and serological techniques are available for the diagnosis of amoebiasis this can almost invariably be made by microscopic examination of specimens of faeces for trophozoites or cysts either in wet films or stained preparations; similarly a search can be made for trophozoites in pus from abscesses or material obtained by biopsy.

Faeces must be examined within an hour of the specimen being passed but if this is impossible then the specimen should be fixed in 4 or 5 times its own volume of polyvinyl alcohol or placed in a refrigerator; on no account should a 'delayed' specimen be kept warm since the amoebae will disintegrate rapidly. Naked eye examination of the specimen should not be ignored and any visible mucus should also be examined microscopically since amoebae usually adhere to it.

AMOEBIASIS

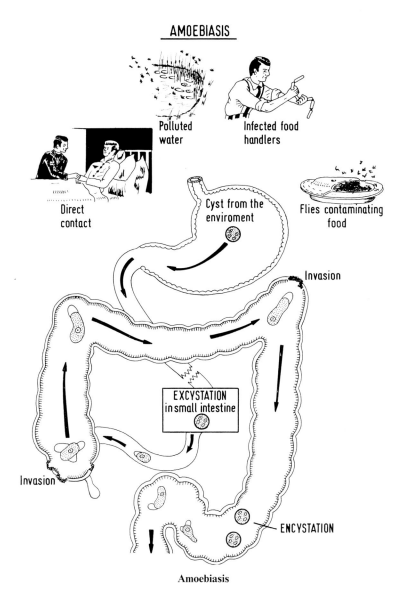

Polluted water

Infected food handlers

Direct contact

Cyst from the enviroment

Flies contaminating food

Invasion

EXCYSTATION in small intestine

Invasion

ENCYSTATION

Amoebiasis

The trophozoite form of *Entamoeba histolytica* is responsible for infection within the human host and the cyst form, which is very much more resistant to adverse environmental conditions, transmits infection from man to man by direct contact with a carrier or case of amoebic dysentery. Indirect spread via polluted water supplies or food infected from food handlers who are carriers or by flies contaminating food after feeding on faeces from a carrier or case also occurs.

When preparing a sample for microscopy the microscope slide and the saline used for mixing the specimen must be heated to body temperature since amoebae become sluggish or immobile in a cold environment. A small portion of the specimen is mixed with warm saline on a warmed slide and *the preparation should not be too thick* or the amoebae may be obscured by debris; the preparation is best viewed by phase contrast microscopy and the presence of amoebae containing red blood cells allows the clinical diagnosis to be clinched with confidence.

When a search is to be made for cysts it is customary to concentrate these from a sample of faeces by the zinc sulphate flotation technique which allows the cysts to float whilst most of the faecal matter sediments.

CONCENTRATION METHOD
FOR CYSTS

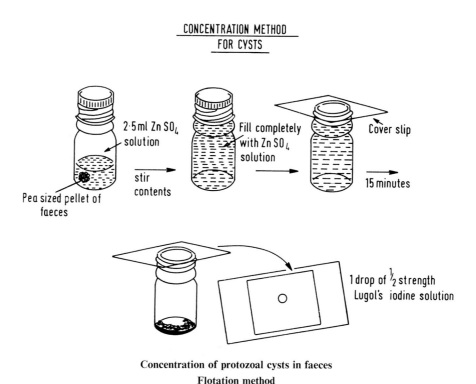

Concentration of protozoal cysts in faeces

Flotation method

A 33 per cent. solution of zinc sulphate ($ZnSO_4$) is used and 2·5 ml. are placed in a 5 ml. screwcap bottle; a sample of faeces is thoroughly emulsified in the solution and then the bottle is filled to the brim with $ZnSO_4$. Any large particles which float to the surface are removed and the solution replenished to be level with the top of the bottle; a coverslip is then laid in contact with the fluid surface.

After 15 min. the coverslip is removed and placed, wet surface downwards, on a microscope slide on which a drop of $\frac{1}{2}$ strength Lugol's iodine has been deposited.

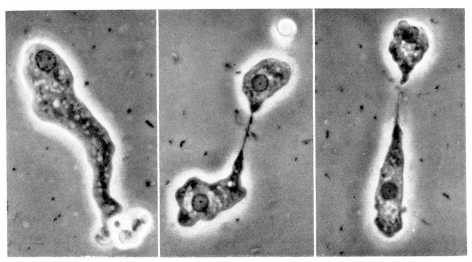

Entamoeba histolytica. Time lapse photography under phase contrast showing a trophozoite dividing; the character of the nuclei should be noted, particularly the ring of fine chromatin granules lying on the nuclear membrane and the small karyosome which is situated centrally. These features which are usually seen even more clearly in stained preparations allow differentiation from the commensal *Entamoeba coli* where the karyosome is not centrally situated and the chromatin is coarser. ×750

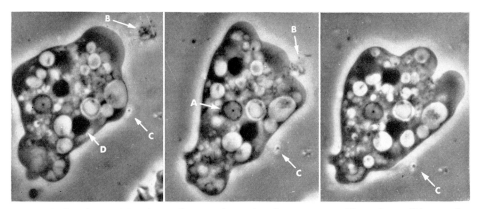

Entamoeba histolytica. Time lapse photography under phase contrast. The point at C is a marker against which the movement of the trophozoite can be judged; at point B are particles towards which the trophozoite progresses and eventually engulfs. D indicates a recently ingested red blood cell and at A the characteristic nucleus is evident. ×750

The cells in both of the above series were harvested from a culture in Jones' medium.

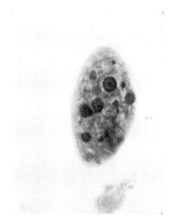

Entamoeba histolytica. Trophozoite in a film of stool stained by haematoxylin and alcoholic eosin; the nuclear characteristics noted in the phase contrast films are also evident here. Red blood cells in various stages of digestion can also be seen. ×750

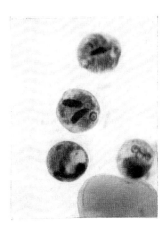

Entamoeba histolytica. Film of faeces from a carrier after concentration of the specimen by the zinc flotation technique and staining with haematoxylin and alcoholic eosin. Four cysts of *Ent. histolytica* are seen and these show variously the deeply stained chromatoid bodies and also nuclei (two or four). ×750

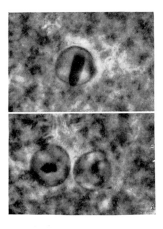

Entamoeba histolytica. A phase contrast view of the cystic stage; these are spherical with a thin smooth wall containing a homogeneous cytoplasm within which the chromatoid bodies show as dark objects. ×750

CHAPTER 38

GIARDIA LAMBLIA

THIS flagellate protozoon inhabits the duodenum and jejunum and although it can be detected in the faeces of healthy individuals it is also incriminated as a cause of epidemic enteritis in young children, particularly those living in residential or day nurseries.

G. lamblia reproduces by binary fission and exists both as a trophozoite and in a cyst form.

The *trophozoite* is flat and pear-shaped when viewed dorsally and in the lateral view is crescent-shaped, the concave ventral surface acting as a sucker to fasten the organism to the intestinal epithelium; it measures approximately 10 by 15µ. Two nuclei are present near the broader anterior end and a comma-shaped body, the median body, lies posterior to the nuclei. Four pairs of flagella originate on the ventral surface anteriorly and the trophozoite has a characteristic rolling movement when seen in wet preparations.

Cyst forms are oval and somewhat smaller than the trophozoite and possess two or four nuclei; parallel axostyles may also be noted.

Giardia species which are found in the intestine of other mammals are morphologically similar to *G. lamblia* but are given specific status according to the host which is parasitised, e.g., *G. canis* in dogs; it is not known whether transmission from animal to human host, or vice versa, occurs.

Laboratory diagnosis depends on the microscopic detection of trophozoites AND/OR cysts in faecal specimens; the zinc sulphate flotation method for cysts (p. 192) can also be used.

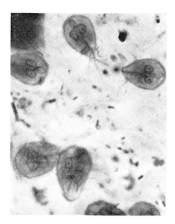

Giardia lamblia from the faeces of a case of giardiasis in a residential nursery; two nuclei are clearly seen near the broader anterior end of each of the five trophozoites; the flagella are not very distinct but the median bodies are seen in the lower two cells. Basic fuchsin ×1000

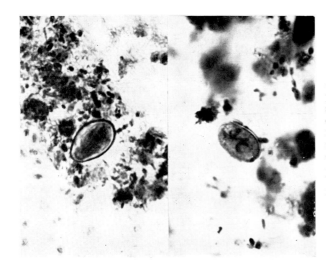

Giardia lamblia. These two pre-parations were made from faecal specimens obtained from two children who had recently suffered from giardiasis. Both show a single cyst form of the parasite each of which contains two organisms formed by sub-division. The parallel axostyles can be noted in the cyst on the right. ×1000

TRICHOMONAS VAGINALIS

Trichomonas vaginalis is the only member of the genus which is pathogenic to mankind although others are common parasites in the intestinal tract of many vertebrates and one other, *T. foetus*, infects cattle and causes early abortion.

T. vaginalis is ovoid, measuring 10 by 15μ and possesses four anterior flagella and an undulating membrane which extends half way down one side of the organism; the single nucleus is situated near the broader anterior end. *T. vaginalis* does not form cysts but the trophozoites can survive outside the host for a day or two provided that the environment is suitable, e.g., cool and moist.

A similar or perhaps identical trichomonas has been isolated from urethral and prostatic secretions in men and from the faeces of both sexes; there is, however, some doubt as to their pathogenicity in these situations but there is no doubt that *T. vaginalis* causes vaginitis which may occur as an entity or in association with gonococcal infection.

Laboratory diagnosis. This is usually undertaken by microscopic examination of wet films of secretions but *T. vaginalis* can also be cultured in suitable media.

197

o

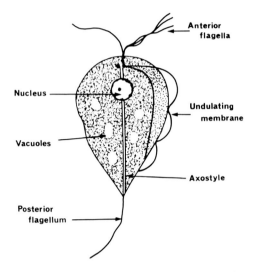

Line drawing showing the essential features of trichomonas species.

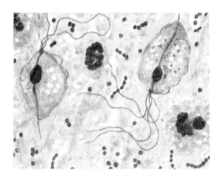

Trichomonas vaginalis from a vaginal smear stained with Leishman's stain. The anterior flagella and the undulating membrane are clearly visible as are the anteriorly situated nucleus and the central axostyle. Pus cells and short chains of cocci (? enterococci) are also present. ×1000

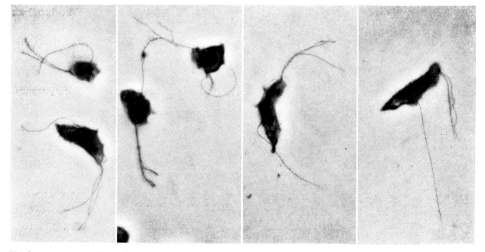

Trichomonas vaginalis. Time lapse photography under phase contrast of *T. vaginalis* from a culture. The characteristically jerky movements of the organism and rotation on its long axis make it difficult to record certain characteristics in wet preparations although the undulating membrane can be noted in the lower organism in the first frame and also in that in the third frame. At least two and sometimes three of the four anterior flagella can be noted in all of the organisms. ×1000

CHAPTER 39

TOXOPLASMA GONDII

Toxoplasma gondii, the causal organism of toxoplasmosis, obtained its specific name from the fact that the disease was first noted in the gondi, an African rodent, at the beginning of this century; more than thirty years elapsed before toxoplasmosis was recognised as occurring in man. *T. gondii* is now known to parasitise many species of mammals and birds and is exceptional amongst protozoa in being able to develop in all kinds of host cells with the exception of the non-nucleated red blood cell.

The trophozoites of *T. gondii* are crescent-shaped, 6 by 2μ, with a single nucleus and in the early stages of infection trophozoites multiply rapidly within the host's cells; trophozoites reproduce by endodyogeny, i.e., schizogony but with the formation of only two daughter cells within the parent. Later in the clinical illness only 'cysts' are found and again these occur in any tissue but particularly in the brain; the 'cysts' are spherical and may reach a diameter of 60μ or more and contain several hundred trophozoites.

Recent experimental studies have shown that in the cat, *T. gondii* can form oval oocysts (9 by 12μ) which are excreted in the faeces and on ingestion can give rise to infection in cats; it may be that this mode of infection also occurs in other animals and in man. Heretofore it has been assumed that infection in man takes place by contact with the large 'cysts'. Transplacental infection of the foetus, often from an asymptomatic mother, can also occur with the prospect of abortion or the child being born with deformities.

It must be stressed however that clinically apparent infection in man is a rare event.

Laboratory diagnosis is by microscopic examination of smears and sections of biopsy material for trophozoites and large 'cysts'; on occasion such material can be inoculated into mice which are killed six weeks later when the mouse brain is sectioned and examined for 'cysts'.

Sero-diagnostic tests are available, e.g., the Sabin-Feldman dye test, complement-fixation and immunofluorescent techniques and much of the epidemiology of toxoplasmosis was elucidated by surveys using the dye test.

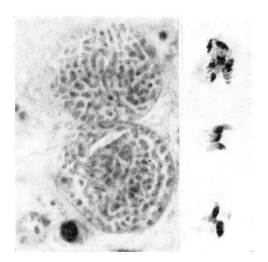

Toxoplasma gondii. On the left is a section of mouse brain from an experimentally infected animal; two 'cysts' are shown and each contains numerous trophozoites. On the right are films made from the peritoneal exudate of the same mouse; free trophozoites are seen and these tend to show a crescent shape with a single central nucleus. Haematoxylin and Eosin. Brain section ×1200. Peritoneal exudate ×1200

CHAPTER 40

TRYPANOSOMES

TRYPANOSOMES are parasitic flagellated protozoa and those species which parasitise vertebrate hosts require a vector for transmission between hosts. Mammalian trypanosomes are either *stercorarian* or *salivarian* depending on the location in the insect vector of the metacyclic forms, i.e., the infective forms which appear at the end of the life cycle in the insect; stercorarian trypanosomes are those in which the metacyclic forms occupy a posterior site in the gut and are excreted in the insect's faeces. The metacyclic forms of salivarian trypanosomes are in the insect's proboscis and are injected into the host when the vector feeds.

Although most mammalian trypanosomes are stercorarian they display a high degree of host specificity and are not pathogenic but one species, *Trypanosoma cruzi* is exceptional in being pathogenic and for a wide variety of hosts.

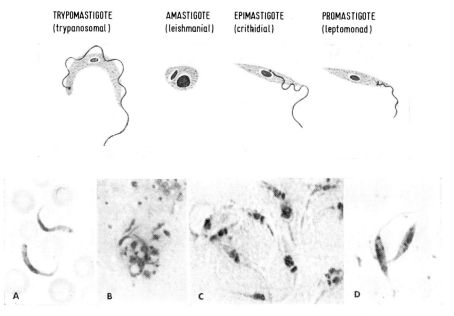

Trypanosomes do not reproduce sexually and commonly multiply by binary fission; they are pleomorphic in that the form of the cell varies during the life cycle and the following terms describe the different forms.

Trypomastigote (trypanosomal form) when the kinetoplast is situated posterior to the nucleus.

Amastigote (leishmanial form) does not possess a flagellum.

Epimastigote (crithidial form) when the kinetoplast is adjacent to the nucleus.

Promastigote (leptomonad form) in which the kinetoplast is near the anterior of the cell.

Stained preparations from blood films (A—D) are shown below the respective diagrammatic representations.

203

TRYPANOSOMA CRUZI

This species is widely distributed throughout South America and the southern states of the U.S.A. In addition to man, in whom it causes Chagas' disease, it parasitises a wide range of wild and domestic animals. The vectors are triatomid bugs—cone-nosed bugs, and Chagas' disease is frequently acquired by those living in bug-infested dwellings; the disease is frequently zoonotic in origin but human-to-human infection via the vector does occur.

Metacyclic trypanosomes are deposited on the skin of the host since the bug usually defaecates whilst it is feeding and the infective forms may enter the host via the puncture wound; however, since the bugs feed at night and often on the host's face the metacyclic forms can be spread over the face and enter the host's conjunctivae, nose or mouth.

The *trypomastigote* form then circulates in the bloodstream before invading tissue cells, particularly muscle; in the tissues the trypanosome changes to the *amastigote* form and reproduces by binary fission with the formation of a 'colony' containing many hundreds of amastigotes.

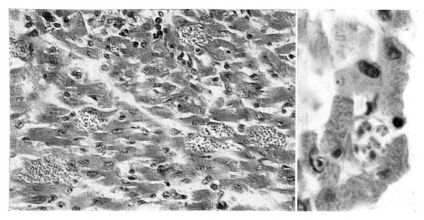

Trypanosoma cruzi. Sections of heart muscle of mouse infected with *T. cruzi.* On the left (×250) at least nine distinct clusters of parasites, embedded in the muscle fibres, can be seen as collections of blue dots; on the right (×1000) and immediately below centre the nuclei and kinetoplasts of the amastigote form of *T. cruzi* can be seen. Haematoxylin and Eosin.

After reproduction in the tissues the cells revert to the trypomastigote form and again invade the bloodstream; thus alternating cycles of reproduction in the tissues and invasion of the bloodstream occur. As infection continues the trypomastigotes become increasingly rare.

When a cone-nosed bug feeds on a patient who has trypomastigotes circulating in his blood, the trypomastigotes then develop in the insect's midgut and *epimastigote* forms appear which spread to the hind gut and within a week or so metacyclic forms are excreted by the bug.

CHAGA'S DISEASE

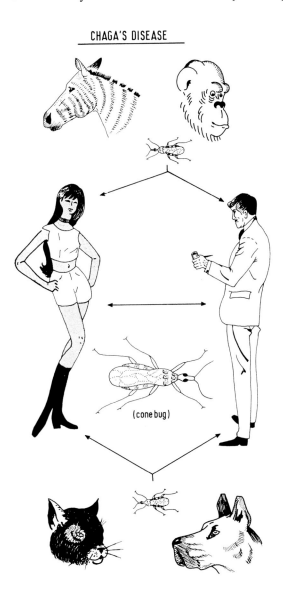

(cone bug)

Chagas' disease

This infection is caused by *Trypanosoma cruzi* which parasitises a wide range of wild and domestic animals whence it is transmitted to man by triatomid (cone-nosed) bugs; human-to-human infection by the vector can also take place.

Laboratory diagnosis can be attempted by microscopic examination of blood films taken during the early, acute stage of infection; *T. cruzi* in the trypomastigote form in the blood has a characteristic appearance, some 20µ in length and with a curved, often C-shaped, appearance. The centrally placed nucleus stains darkly and an even more darkly stained kinetoplast lies in the posterior part of the cell; the undulating membrane associated with the flagellum is often seen. The flagellum arises from the vicinity of the kinetoplast and passes along the cell surface to point anteriorly.

Xenodiagnosis may also be undertaken, i.e., laboratory-reared bugs are allowed to feed on the suspected case and after some days the bugs' faeces are examined for trypanosomes.

In chronic infection serological diagnosis must be relied on and although complement-fixation tests with patient's serum have been widely used in the past, haemagglutination and fluorescent antibody techniques are now available and are much more sensitive.

Of the several non-pathogenic stercorarian trypanosomes only *T. rangeli* is likely to cause confusion in the laboratory diagnosis of Chagas' disease since it has the same host range and vectors as *T. cruzi* and occurs in the same parts of the world; the two species can be easily differentiated in blood films after a little experience.

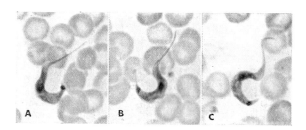

Trypanosoma cruzi. The characteristic curved shape is evident in these three blood films as is the darkly stained posteriorly sited kinetoplast. Thick blood films are more satisfactory for diagnosis since in thin films the trypanosomes frequently break up with loss of much of the cytoplasm.
×1200

TRYPANOSOMA GAMBIENSE

This is one of the two salivarian trypanosomes which can infect man, the other is *T. rhodesiense;* their morphology is identical as is their development in the vector, the tse-tse fly.

These trypanosomes are found in tropical Africa within the latitudes 15°N and 30°S but whereas *T. gambiense* occurs in West Africa and Uganda and causes a chronic form of sleeping sickness lasting months or years, *T. rhodesiense* gives rise to acute sleeping sickness in East Africa; the latter trypanosome has a wide host range and is usually a zoonotic infection whereas animals are not a significant source of infection with *T. gambiense* which for all practical purposes is a human parasite.

In the infected patient the trypanosomes reproduce by longitudinal binary fission in the blood and other body fluids and recurring parasitaemia is a feature of the early stages of the illness; the organisms can also be detected in enlarged lymph glands and later in the disease when trypanosomes are scanty or absent from the bloodstream they can be found in the cerebrospinal fluid.

Development of either species within the tse-tse fly is identical but much more complex than in the human host; trypanosomes ingested by the fly pass to the midgut where within the peritrophic membrane they assume a long broad form. They escape at the posterior end of the midgut to lie between the membrane and the gut wall and a week or ten days after being ingested the parasites are found closely packed in the extra-peritrophic space and here they have become long and slender; they must then penetrate the peritrophic membrane and reach the main food-channel to migrate along the foregut to the proboscis. Thereafter they pass up the hypopharynx to the salivary glands where they change into epimastigote forms; metacyclic trypanosomes develop from the epimastigotes and pass with the saliva down the hypopharynx to be inoculated when the fly bites a new host. Throughout the development in the tse-tse fly reproduction takes place by binary fission.

Laboratory diagnosis depends on the microscopic detection of the parasite in blood films, aspirated lymph gland juice or cerebrospinal fluid. Inoculation of such material intraperitoneally into mice or rats is also an efficient detector of *T. rhodesiense* (these animals are not susceptible to infection with *T. gambiense*); tail blood from the inoculated animal is examined for the presence of trypanosomes from the third day onward.

Sero-diagnostic procedures are not presently useful because of the antigenic variability of trypanosomes.

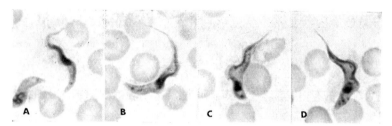

Trypanosoma gambiense. Morphologically *T. gambiense* is identical with *T. rhodesiense* and *T. brucei*. The undulating membrane is seen in B and C and the small subterminal kinetoplast is not as conspicuous as in *T. cruzi*. The pleomorphic nature of *T. gambiense* is exemplified by comparing the stumpy form in A (lower left) which has no free flagellum with the slender forms in B and C and the intermediate form in D where the nucleus is displaced to the posterior part of the cell and the flagellum is short. ×1200

OTHER TRYPANOSOMES

Many other trypanosomes are associated with diseases in animals, e.g., *T. brucei*, which has a wide range of hosts some of which suffer severe and frequently fatal infections whereas others display little or no effect from the parasitisation; with the exception of infection with *T. equiperdum* which is transmitted between horse or donkey hosts by sexual intercourse, all of these animal trypanosomes require an insect vector.

LEISHMANIAE

Various species of *Leishmania* parasitise man and other mammals but only man suffers disease; transmission is by sandflies and is frequently zoonotic in origin. Leishmaniae differ from trypanosomes in that their development in the mammalian host is solely intracellular.

LEISHMANIA DONOVANI

L. donovani causes kala-azar (visceral leishmaniasis) in man and also parasitises dogs and wild animals; thus dogs, which themselves show no evidence of parasitisation, act as reservoirs whence sandflies can transmit the leishmania to man.

L. donovani introduced to a host by the sandfly's bite first proliferate at the site of inoculation and then spread by the lymphatics and bloodstream to all organs possessing reticulo-endothelial cells; in the host the promastigote forms introduced by the sandfly rapidly change to the amastigote form and are ingested by macrophages within which they divide by binary fission until the host cell is crammed with parasites. When the host cell ruptures the amastigote forms are liberated to be taken up by fresh macrophages.

If a sandfly takes a blood meal from a host when infected macrophages are available, some of these will be ingested and the amastigotes are released from the macrophages in the insect's midgut and change to the promastigote forms; these multiply rapidly and migrate to the foregut ready to be inoculated into a fresh host; from ingestion of infected macrophages by the sandfly an interval of 5-7 days elapses before promastigotes appear in the foregut.

The amastigote form which infects man is an ovoid cell, some 2 to 4µ in size and does not possess flagella; each cell possesses a nucleus and a kinetoplast and cells in this form are referred to as Leishman-Donovan (LD) bodies, and are diagnostic of leishmaniasis.

Laboratory diagnosis. In addition to microscopic examination of specimens of lymph gland, bone marrow or spleen for LD bodies, samples of such biopsy materials or blood may be inoculated intradermally or intraperitoneally into a hamster which is highly susceptible to leishmanial infection; biopsy of skin or spleen from the hamster is undertaken some weeks later and stained sections are examined for LD bodies.

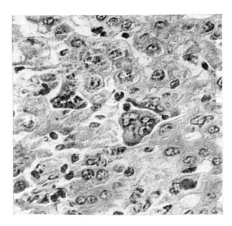

Leishmania donovani. Section of spleen showing Leishman-Donovan bodies; these comprise numerous amastigote forms of the parasite within the host cells. × 500 Haematoxylin and Eosin

LEISHMANIA TROPICA

L. tropica is identical with *L. donovani* in morphology and its development both in the insect vector and mammalian hosts; however, it causes lesions only in the skin, i.e., cutaneous leishmaniasis, oriental sore or Delhi boil. Diagnosis is by microscopic examination of material obtained from the edge of the skin ulcer.

Cutaneous leishmaniasis in man in the Americas is not usually a solitary lesion and metastatic spread occurs, e.g., in Brazil the disease, espundia, caused by *L. brasiliensis* affects the mucosa of the nasopharynx and also the nasal cartilage. *L. brasiliensis* is identical in morphology and life cycle with *L. donovani.*

CHAPTER 41

MALARIA

MAN is the intermediate host for at least four species of plasmodia, *Plasmodium falciparum*, *P. malariae*, *P. ovale* and *P. vivax*; two other species, *P. cynomolgi* and *P. knowlesii*, whose natural intermediate hosts are simians can also be transmitted to man at least experimentally.

The definitive hosts for these plasmodia are various species of anopheline mosquito which do not suffer any ill effect from parasitisation; contrarily man suffers malaria which, in the absence of specific therapy, is characterised by regularly recurring attacks of severe fever. Malaria occurs extensively throughout those parts of the world where conditions are suitable for breeding of the relevant mosquito and where they have access to the human host.

Malaria is one of the most important infections in the world today, both in regard to morbidity and mortality and its effect on the economy of countries such as India must be tremendous although control measures have effected its eradication from certain areas. The complete life history of malarial parasites was elucidated little more than twenty years ago when experiments with monkey hosts revealed that injected parasites migrated rapidly to the parenchyma cells of the liver; subsequently a human volunteer was infected and subsequent liver biopsy showed that a pre-erythrocytic cycle also occurs in man.

That part of the life cycle of malaria plasmodia which occurs in the human host is known as schizogony and is asexual although gametocytes are formed in the host; the sexual cycle or sporogony takes place in the female anopheline mosquito.

SCHIZOGONY

After the human host has been bitten by an infected female mosquito the sporozoites thus introduced leave the general circulation within a short time and invade tissue cells with a predilection for the parenchyma cells of the liver; within these latter cells the sporozoites divide asexually to form schizonts which ultimately occupy the entire liver cell and are teeming with thousands of merozoites. At the end of this pre-erythrocytic cycle which lasts 7 to 10 days, the infected liver cells rupture and the liberated merozoites invade red blood cells in the general circulation.

This pre-erythrocytic cycle lasts for different periods of time depending on the species of plasmodium causing the infection and with the exception of *P. falciparum* infection, the merozoites released from the hepatic schizonts also attack other liver cells so that an exo-erythrocytic cycle continues in the liver in parallel with the erythrocytic cycle.

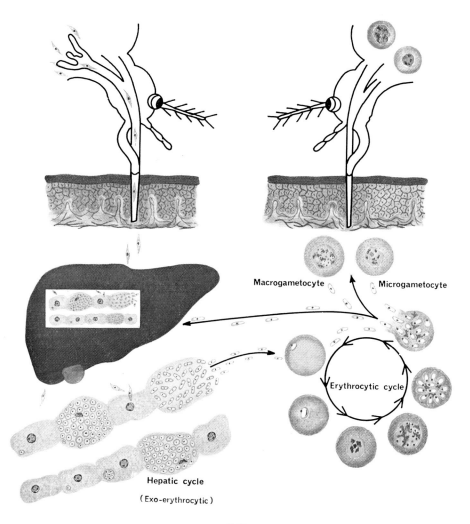

Macrogametocyte

Microgametocyte

Erythrocytic cycle

Hepatic cycle

(Exo-erythrocytic)

Schizogony

This diagram summarises the sequence of events from the time that sporozoites are inoculated from the mosquito on the left into man, through the exo-erythrocytic hepatic cycle, then the erythrocytic cycle until another mosquito (right) takes a blood meal from the patient and ingests gametocytes along with merozoites. The 'window' in the liver is shown in enlarged form immediately below that organ and depicts the development of schizonts and the release of merozoites when the liver cells rupture; similarly the progression of the erythrocytic cycle from the invading ring form through the formation of trophozoites to schizonts is shown. The merozoites liberated from erythrocytic schizonts can then invade fresh red blood cells, healthy liver parenchyma cells or be ingested by another mosquito along with gametocytes.

Schizogony is thus a continuing process in the liver and red blood cells and excepting infection with *P. falciparum* the persistence of an exo-erythrocytic, hepatic cycle gives rise to relapses of the infection.

The duration of the erythrocytic cycle also varies depending on the parasite involved and is determined by the interval required for the merozoite infecting a red cell to pass from the ring-form stage to become a trophozoite and in turn a schizont with the ultimate rupture of the red cell and the liberation of new merozoites; the interval is 48 hours in the case of *P. ovale* and *P. vivax*, 36 to 48 hours for *P. falciparum* and 72 hours for *P. malariae*.

Thus classically in a human host bitten on only one occasion and infected with only one species, schizogony will be synchronous since virtually all of the parasites will mature simultaneously with massive numbers of merozoites being liberated in a short space of time so that febrile paroxysms will occur every second day in the case of infection with *P. vivax* or *P. ovale* and every third day with *P. malariae* infections. This periodicity is influenced however not only by therapy but also if the patient suffers multiple infection with the same species or where more than one species infects the same host.

During the erythrocytic cycle a few parasites become differentiated as either male (microgametocyte) or female (macrogametocyte) cells so that the sexual cycle commences in the human host.

SPOROGONY

Before sporogony can be fulfilled the male and female gametocytes must be ingested by a female mosquito and further development of the gametocytes takes place in the stomach or midgut of the vector once the gametocytes have escaped from the ingested red cells. The macrogametocyte matures to become a macrogamete by forming one or two polar bodies which become detached; the macrogamete is then ready for fertilisation.

Maturation of the microgametocyte is known as exflagellation and is a somewhat more complex process; firstly the nucleus of the microgametocyte divides three times and the resultant eight nuclei move to the periphery of the cell and each then moves into separate flagellar-like structures which have formed at the cell surface; the microgamete is then ready to fertilise a macro-gamete with the formation of a zygote. The zygote in turn becomes an ookinete, a crescent-shaped motile cell which migrates through the gut epithelium within 24 hours or so of being formed. Further development takes place on the outer surface of the stomach under the basement membrane; an oocyst is formed by rounding up of the ookinete and the development of an encysting membrane and within the growing oocyst the nucleus divides repeatedly and ultimately the cytoplasm divides to form hundreds of sporozoites. These infective forms of the parasite are liberated into the coelom whence they migrate to the salivary glands of the mosquito to await injection into another host.

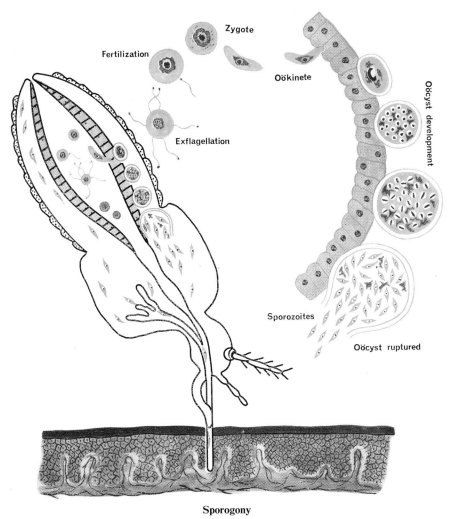

Sporogony

This diagram outlines the progression of events in the sexual stage of the life cycle of malaria parasites in the mosquito; two gametocytes have been ingested and after maturing to macrogamete and microgamete—the latter showing exflagellation, fertilization occurs with the formation of a zygote which develops into the motile, crescent-shaped ookinete which then penetrates the gut wall to lie immediately under the basement membrane where the oocyst forms with the eventual liberation of sporozoites into the coelom. The sporozoites migrate to the salivary glands and await injection into a human host when the mosquito next takes a blood meal.

Laboratory diagnosis. This is based on the microscopic examination of stained blood films; thick films are used to scan for the presence of malarial parasites and identification of the species infecting the patient is based on examination of thin films.

Species identification rests on the differing morphology of the erythrocytic forms of the parasites and some of these are exemplified photographically.

A thin blood film is prepared by collecting on the narrow edge of a microscope slide a drop of blood about the size of a pin head; this spreader slide is then applied at an angle of 30° to a second slide on a horizontal surface and pushed along so that the drop of *blood is trailed behind* the interface of the slides; if the blood is pushed instead of being dragged then the malarial parasites may be crushed.

A thick blood film is prepared by collecting in the centre of a slide a drop of blood two or three times the volume of that required in preparing a thin film; the drop is then spread in a circle of 1 to 2 cm. diameter using the corner of another slide. Blood films are allowed to dry in air before being stained by Leishman's or some other similar stain.

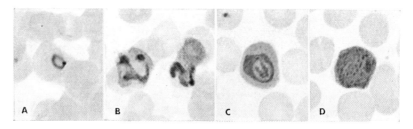

Plasmodium vivax

At A the merozoite has assumed a ring form soon after entering a red cell; a rim of cytoplasm surrounds the central vacuole and the nucleus lies within the cytoplasm—the nucleus is referred to as the chromatin dot.

In B are developing trophozoites which arise from the ring form within a few hours; the vacuoles are reduced in size and the parasites are irregularly shaped.

C shows a microgametocyte with the nucleus centrally placed and comprising a diffuse skein of fibrils.

D. A macrogametocyte; this female form is larger than the microgametocyte and expands the red cell. The nucleus is characteristically compact and sited peripherally, in this instance at the right side of the parasite. ×1200

It should be noted that the gametocytes of *P. malariae* and *P. ovale* are similar to those of *P. vivax* in their general characteristics but are not so frequently seen in peripheral blood. Those of *P. malariae* are smaller than *P. vivax* whereas the gametocytes of *P. ovale* are intermediate in size.

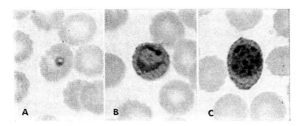

Plasmodium ovale

A. A ring form which is thicker than the delicate ring form of
P. vivax.

B shows a developing trophozoite which is very much more
compact than that of *P. vivax* and at C is a schizont showing
11 or 12 merozoites with pigment aggregated centrally.
×1200

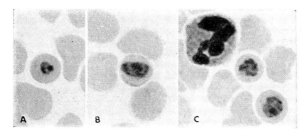

Plasmodium malariae

A shows a ring form and at B a compact band form of the
developing trophozoite; this form is common in *P. malariae*
infections.

Two immature schizonts are seen at C—centre and bottom
right. ×1200

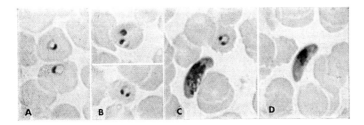

Plasmodium falciparum

A shows red cells and two are infected with ring forms; in the upper part
of B one red cell is infected with two ring forms and in the lower part
another characteristic of *P. falciparum* ring forms may be noted,
i.e., that there are two chromatin dots in the parasite.

C shows the features of the microgametocyte, i.e., diffuse scattering of
chromatin granules and pigment; a ring form with two chromatin dots
is also evident. Finally the compact masses of chromatin and pigment
near the centre of the macrogametocyte are shown in D. ×1200

SECTION V

MYCOLOGY

CHAPTER 42

CLASSIFICATION AND TECHNIQUES

THERE are thousands of species of fungi and as with bacteria the vast majority are not parasitic on man or animals but follow a saphrophytic existence; indeed many species are helpful to man by participating in the decomposition of animal and plant debris. Others, e.g., *Penicillium notatum* produce antibiotic substances used in the treatment of bacterial infections.

Many saprophytic species annoy bacteriologists by contaminating laboratory culture media and also irritate food producers and housewives by causing spoilage of foodstuffs. Apart from causing infection in man and animals the spores of certain fungi, e.g., *Aspergillus clavatus*, are incriminated in allergic reactions.

CLASSIFICATION OF FUNGI

Fungi pathogenic to man can be classified on a morphological basis into four groups:

Moulds which grow as branching filaments (hyphae) that interlace and form a dense, felted mass (mycelium); such filamentous fungi have a vegetative mycelium which grows into the substrate and absorbs nutrients whereas the aerial mycelium rises from the surface of the mould and allows dissemination of sexual spores.

Yeasts are unicellular fungi appearing as single round or oval cells; reproduction is by budding from the parent cell and sexual spores are not formed. When grown on solid media yeasts form colonies similar to those of bacteria in contrast to the powdery colonies of moulds.

Yeast-like fungi grow either as round or ovoid cells or as non-branching filaments; like yeasts they also reproduce by budding and on solid media produce colonies similar to those of staphylococci.

Dimorphic fungi comprise the fourth morphological group and are so called since when growing in tissues or *in vitro* at 37°C they appear in yeast forms whereas when cultures are incubated at 22°C they present as mycelial growth.

A more formal and systematic classification of fungi exists which is primarily dependent on the nature of their sexual processes; since these are difficult to induce the formal classification is not offered here but it should be noted that most of the fungi pathogenic to man fall into Class IV of that classification, i.e., the *Fungi imperfecti*, a class which groups together all fungi which do not have a sexual stage or where such a stage has not yet been demonstrated.

TECHNIQUES IN MYCOLOGY

Microscopy. Microscopic examination of unstained or stained material plays a large part in the identification of fungi; indeed microscopy of the specimen from the patient and from resultant cultures is frequently the only manoeuvre required to arrive at a diagnosis.

Specimens of skin scales, hair or nail clippings from cases of ringworm (tinea) must first be rendered transparent ('cleared') to allow observation of the infecting fungi; fragments of either type of specimen are placed on a glass slide and covered with 20 per cent. sodium or potassium hydroxide and a cover slip is then applied. The preparation is then left at room temperature to allow the keratin to be partially dissolved; skin scales will be cleared within 5-10 min. whereas pieces of nail may have to be treated for an hour or two at 37°C before digestion has reached the stage at which fungi may be seen. In that case the hydroxide must be replenished from time to time or alternatively the specimen may be immersed in the agent in a test tube and incubated before being transferred to a microscope slide for examination.

The cover slip is pressed down gently with blotting paper to give a thin film which is then examined under the dry objectives; cholesterol crystals in skin and oil or fat droplets may be mistaken for fungal elements but this potential source of confusion can be eliminated by replacing the hydroxide with lactophenol blue stain which is not accepted by such artefacts.

Staining. If it is desired to stain a cleared preparation lactophenol blue stain is applied at one edge of the cover slip and the hydroxide is drawn from the opposite edge with blotting paper until the stain has completely replaced the alkali.

Gram's staining method is also used in the diagnosis of fungal infections, e.g., in candida infections, either directly on material from cases or on cultures obtained from specimens; India ink preparations may also be used, e.g., to demonstrate the capsules of *Cryptococcus neoformans* in cerebrospinal fluid.

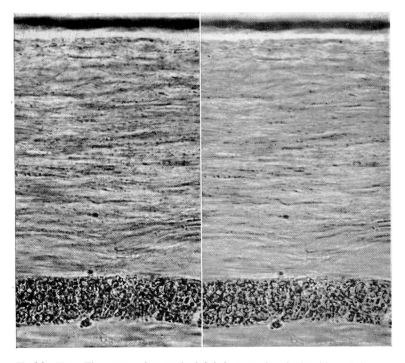

Healthy Hair. The preparation on the left is in potassium hydroxide and that on the right after the hydroxide was replaced with lactophenol blue; slightly more than half the width of the hair is shown with the surface cuticle at the top and the medullary core near the lower edge. The cortex comprises more than two thirds of the hair shaft. This preparation should be compared with those showing hairs infected with ringworm fungi. ×500

Cultivation. The most commonly used medium for diagnostic purposes is Sabouraud's glucose peptone agar which has a pH of 5·4 and cultures may be made in parallel on this medium with or without the addition of thiamine (10 mg. per litre); the latter promotes spore formation by certain fungi causing ringworm. The aerobic nature of fungi must be catered for when attempting their isolation as must the fact that they grow more slowly than bacteria and culture plates should be incubated at 20°C for at least three weeks before being discarded although most positive cultures will show evidence of growth within a few days and be typical for identification by 7 to 14 days.

In the case of ringworm fungi colonial appearances should be noted not only of the surface growth but also on the reverse side of the colony and attention should be paid to any pigment which may have diffused into the medium; the majority of ringworm fungi give large spreading colonies covered by a fluffy or powdery aerial mycelium; the various pigments which are produced may be confined to the colony or diffuse into the medium for a greater or lesser distance. Some colonies are exemplified but no attempt has been made to give any detail of colonial morphology.

Needle Mount Preparations. Microscopic examination of material from colonies allows differentiation of the three genera of fungi responsible for ringworm; a needle mount preparation is obtained by removing a piece of sporing mycelium from a culture on Sabouraud's agar; the material is placed on a slide and is then teased out in a drop of 95 per cent. ethyl alcohol using two straight wires or needles. Just before the alcohol has completely evaporated a drop of lactophenol blue stain is added and a cover slip is applied: the preparation is then left at room temperature for a few minutes to allow the stain to penetrate. Excess stain is then removed by gentle pressure on the preparation through a sheet of blotting paper before viewing with low-power and high-power dry objectives.

Slide culture technique. By subculturing mycelial growth on an agar block held between a microscope slide and a cover slip the arrangement of the growing mycelium and spores can be observed intermittently and undisturbed during growth.

A block of Sabouraud's agar medium, 1 cm. square and 2 mm. thick is placed on a sterile microscope slide and the block covered with a sterile cover slip; using a straight wire the agar block is inoculated at the midpoint of each of the vertical sides with material from a culture. The assembly is then placed in a closed chamber, e.g., a plastic sandwich box, containing several layers of blotting paper soaked in 20 per cent. glycerol to ensure a humid aerated atmosphere. Incubation is at room temperature or preferably at 28°C.

The slide can be removed at intervals and examined microscopically with the dry objectives without disturbing the assembly; as soon as adequate sporing is noted stained preparations can be made from both the slide and the cover slip.

The cover slip is removed and the agar block discarded; one drop of 95 per cent. ethyl alcohol is applied to the slide and to the cover slip and immediately before the alcohol has completely evaporated one drop of lactophenol blue stain is applied to each preparation; these are then covered with a cover slip and slide respectively. The preparations are left at room temperature for some time to allow penetration of the stain before the excess is removed by gentle pressure through a sheet of blotting paper; if permanent preparations are required the edges of the cover slip can be sealed to the slide, e.g., with nail varnish.

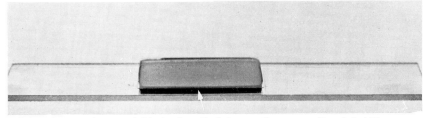

Slide Culture Technique. A block of sterile Sabouraud's agar medium is held between a sterile glass slide and cover slip. The medium is inoculated at the midpoint of each of the vertical sides, e.g., as exemplified by the arrow point.

219

CHAPTER 43

DERMATOPHYTIC FUNGI

THREE genera of the *Fungi imperfecti* namely *Microsporum*, *Trichophyton* and *Epidermophyton* provide the species causing the most common type of fungal infection throughout the world without discrimination between sexes, race or climate; the dermatophytoses, commonly called tinea or ringworm, are superficial infections of the skin, hair and nails which never spread deeper than the keratinous layer but spread peripherally from foci to produce ring-like lesions. Species of ringworm fungi which primarily infect man are *anthrophilic* and these usually produce chronic mild and often subclinical infections although scalp infection with *T. schoenleinii* is exceptional and causes favus; this infection is sited in hair follicles and gives rise to small red papules which develop into scutula, i.e., cup-shaped, sulphur-yellow crusts which eventually destroy the affected hairs.

Among the anthrophilic species some, e.g., *T. mentagrophytes* and *T. tonsurans* are catholic in their habits and can attack nails and the skin and hair anywhere on the body whereas other species are more restricted in the sites where they cause disease, e.g., *E. floccosum* which is most commonly incriminated in tinea cruris and can also infect nails and other skin areas but does not cause infection of the scalp. Similarly *Microsporum* species, commonly the cause of scalp ringworm, do not affect nails. *M. audouinii* infections are restricted to children in whom it is the commonest cause of scalp ringworm and such infection usually disappears spontaneously around puberty.

On the other hand *zoophilic* species are parasitic essentially on various animals but can be transmitted to human hosts, especially children, from domestic pets or farm animals; when such zoophilic species as *M. canis*, *M. gypseum* and *T. equinum* cause human infection they usually cause a more acute inflammatory response than do anthrophilic species.

Laboratory diagnosis of ringworm. As with specimens for bacteriological examination, material submitted for laboratory diagnosis must be taken carefully. Cleansing of the lesion with 70 per cent. alcohol will reduce the chance of bacterial contamination; skin scales should be harvested by scraping the skin at the advancing edge of the lesion with a blunt scalpel. Infected nails are clipped for examination and scrapings are also taken from the deeper parts using a blunt scalpel. Great care is required in selecting hairs for submission to the laboratory; certain species cause the infected hairs to fluoresce under ultra-violet irradiation; for this purpose a Wood's lamp is used to view the head in a darkened room. The stumps of broken hairs should be removed with fine forceps and lustreless hairs from the vicinity of the lesion can also be sent for examination. All specimens are submitted in individual paper envelopes clearly marked for identification of the source.

Laboratory diagnosis is normally by microscopic examination of the infected material after it has been cleared with alkali; microscopic examination of affected hairs shows whether infection is endothrix as in the case of favus, or ectothrix; a note should also be made in the case of ectothrix infections whether infection is of the small-spore or large-spore type. Characteristically *M. audouinii* gives small-spore ectothrix infection as does *T. mentagrophytes;* other *Trichophyton* species produce large-spore ectothrix or alternatively endothrix infections.

Cultivation. Since bacteria may also be present on the specimen it is necessary to treat it with 70 per cent. alcohol for 2 to 3 minutes to reduce the bacterial population before inoculating media. The dermatophytic fungi are aerobic so that fragments of hair, skin or nail must be implanted on the surface and partially submerged. If a pure culture of the fungus is required from the resultant growth then subculture should be made from the *edge* of the spreading colony to fresh medium and this will usually avoid carrying over any bacterial contaminants which may be present.

Needle mount preparations from the aerial mycelium, when stained with lactophenol blue, reveal microscopically the size, shape and distribution of micro- and macroconidia.

Microconidia are small, round or oval single-cell structures on spore-bearing hyphae and are characteristically produced by *Trichophyton* species but are not seen in cultures of *Epidermophyton* species; they are seen only occasionally and in small numbers in *Microsporum*.

Similarly, the morphology of macroconidia can be noted and this further characterises the three genera of the dermatophytes:—

Microsporum species produce large, thick-walled, fusiform macroconidia which are divided into numerous cells by transverse septa although the anthrophilic species *M. audouinii* is exceptional in not producing conidia when grown on Sabouraud's medium.

Trichophyton species frequently do not yield macroconidia when grown on Sabouraud's medium but when they occur they are small, thin-walled and cylindrical.

Epidermophyton species produce thin-walled macroconidia which in young cultures are finger-like and later assume a club or pear-shaped appearance.

Microsporum audouinii. Showing a mosaic of small spores on the surface of the hair, ectothrix, and also the broken brush-like free end of the hair. ×350

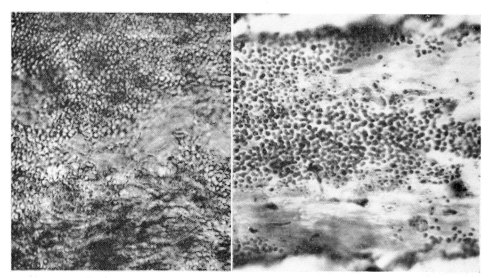

Microsporum audouinii. Hairs from a case of tinea capitis in a child; the preparation on the left is in potassium hydroxide and that on the right after the hydroxide was replaced with lactophenol blue. This is a classical example of small spore ectothrix ringworm infection of hair. ×500

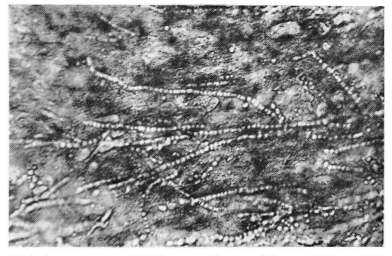

Trichophyton tonsurans. Hair from an early case of tinea capitis in a child; chains of large arthrospores are visible in this endothrix infection. ×500
KOH preparation

222

Trichophyton schoenleinii. Hair from a case of favus showing endothrix infection; a few hyphae can be seen and also empty areas or tunnels where hyphae have degenerated. The presence of air bubbles is distinctive and the absence of the medulla of healthy hair should be noted. ×500 KOH preparation

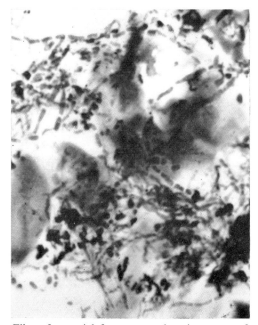

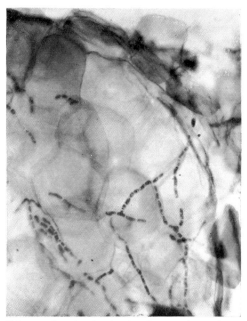

Film of material from a scutulum in a case of favus due to *T. schoenleinii* showing a dense felt work of mycelium and spores with a background of cellular debris. ×500 Lactophenol blue stain

Skin scraping showing hyphae fragmenting to form arthrospores which either by direct contact or indirectly by fomites, e.g., duck boards in spray rooms, can be transmitted to fresh human hosts. ×500 Lactophenol blue stain

Microsporum audouinii. Needle mount preparation from culture on Sabouraud's medium; the virtual absence of micro- and macroconidia is characteristic of this species when grown on Sabouraud's medium and the growth here comprises only sterile interlacing hyphae. ×500 Lactophenol blue stain

Microsporum canis. Needle mount preparation from culture on Sabouraud's medium showing a cluster of large, thick-walled, fusiform macroconidia which are characteristic of most *Microsporum* species. In these mature macroconidia transverse septa have divided each into numerous cells or segments. ×500 Lactophenol blue stain

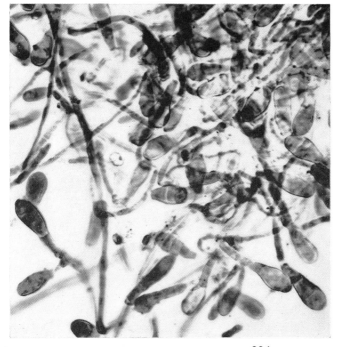

Epidermophyton floccosum. Needle mount preparation from a young culture on Sabouraud's medium showing numerous thin-walled macroconidia which are beginning to assume a club-shaped appearance; the majority are non-septate but a few, e.g., at bottom right corner, show the characteristic formation of 2-4 cells. Microconidia are characteristically absent. ×500 Lactophenol blue stain

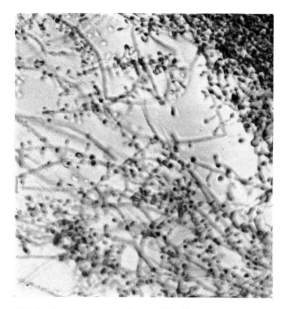

Trichophyton mentagrophytes. Needle mount preparation from culture on Sabouraud's medium showing numerous microconidia; the majority are spherical and borne single from the sides of the hyphae. Macroconidia are absent as is the case with most *Trichophyton* species when grown on this medium. ×500 Lactophenol blue stain

Trichophyton mentagrophytes. This needle mount preparation is exceptional since macroconidia are present in the growth from Sabouraud's medium; six cylindrical, thin-walled macroconidia are present and two of these show septa with division of the macroconidia into segments. ×500 Lactophenol blue stain

225

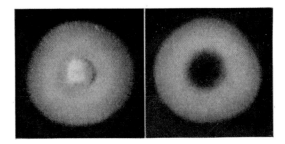

M. audouinii (Growth slow). Flat with sparse short aerial hyphae and creamish white in colour; the reverse is a light tan with a dark salmon-pink centre.

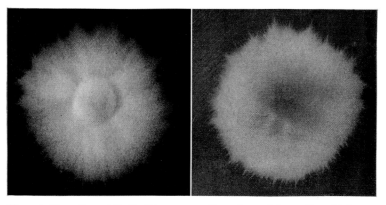

M. canis (Growth rapid). In this young culture the aerial mycelium is characteristically white and fluffy with yellow pigment showing through the peripheral growth; the reverse is characteristically yellow with a hint of the dull orange-brown of an older culture showing centrally.

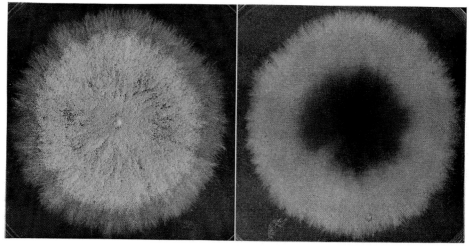

M. gypseum (Growth rapid). Flat with irregularly fringed border and a powdery surface with a greyish-white centre and a cinnamon-brown edge. The reverse is tan with a red central area which is not often seen in this species.

226

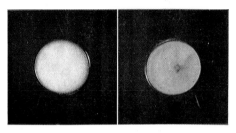

E. floccosum (Growth slow). The white powdery flat surface growth in this relatively young culture becomes tan or olive green later; the reverse is characteristically tan.

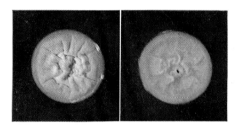

T. schoenleinii (Growth slow). The growth is characteristically irregularly heaped and folded with a waxy cream to yellowish tan colour; the reverse is tan coloured and the colony is somewhat reminiscent of the surface of Camembert cheese.

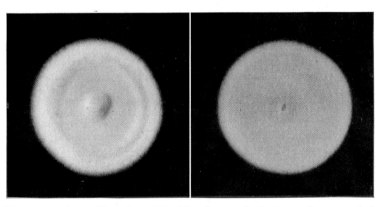

T. mentagrophytes (Growth rapid). The downy, fluffy flat surface growth is white at the periphery and light tan in the older central area; the reverse is tan in this instance but is often colourless and pink or even red.

All of these cultures were grown on Sabouraud's agar for 10 days at 20°C; in each instance the surface growth is shown on the left and the appearance of the base of the colony on the right.

The six plates are intended to exemplify colonial appearances of some ringworm fungi.

MALASSEZIA FURFUR

Malassezia furfur is the cause of a superficial skin infection known as pityriasis versicolor which is worldwide in distribution although it is more commonly seen in hot, humid climates; the infection is almost always asymptomatic and the chronicity of the infection, which produces irregularly shaped brownish desquamating lesions, dictates that it usually presents as a cosmetic problem. It must however be differentiated from more serious infections.

Skin scales from the lesion are examined microscopically and the presence of clusters of oval or round cells measuring 3 to 6μ in diameter together with short, thick hyphal elements approximately 2 to 4μ in diameter is sufficient to confirm the diagnosis.

Cultivation of *M. furfur* from skin scales is of little value since similar organisms can be harvested from normal skin and differentiation of such organisms by biochemical and other methods is not yet valid.

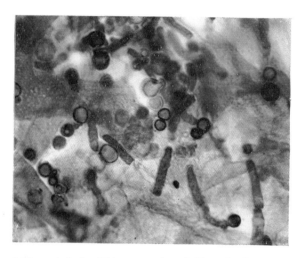

Malassezia furfur. This preparation of skin scales from a case of pityriasis versicolor shows the characteristic round, thick-walled cells, 3-7μ in diameter, and one of these (to the right of centre) is budding; also present are characteristic short, thick hyphal elements. ×1000 Lactophenol blue stain

CHAPTER 44

CANDIDA ALBICANS

ANOTHER fungal infection which is worldwide in distribution is candidiasis (Moniliasis) caused by *Candida albicans* (*Monilia albicans*) or more rarely by other species within the genus which are usually commensal in man and animals. Infection is usually confined to superficial lesions of mucous membranes, particularly in the mouth or vagina, or on the skin.

C. albicans is an opportunist pathogen and the host defence mechanisms have to be weakened by some other process before it can cause infection; oral candidiasis (oral thrush) is most often seen in debilitated new-born babies and vaginal thrush occurs in diabetic women and also not uncommonly in pregnancy when it clears spontaneously after delivery.

Skin infection (cutaneous candidiasis) is usually restricted to flexures and other moist areas of the body but can also occur on the hands of people whose occupation demands prolonged immersion of their hands in water. Diabetics, alcoholics and drug addicts are peculiarly susceptible to infection as are people suffering from leukaemia or iron-deficiency anaemia; similarly infection is promoted if a debilitated host is treated with cortisone or with broad spectrum antibiotics; *C. albicans* is resistant to most antibiotics.

Deep-seated infection, e.g., of the lungs and the intestine, can also occur as can generalised candidiasis and these serious infections are usually associated with the administration of antibiotics for the treatment of bacterial infections or with the use of immuno-suppressive agents.

Laboratory diagnosis. Since *C. albicans* occurs as a commensal at sites where it may declare its opportunistic pathogenic role, the diagnosis of candidiasis in these sites is essentially a clinical one. However, the finding of large numbers of candida in a specimen, especially on repeated occasions, combined with the absence of other pathogens assists in confirming the diagnosis.

Microscopy. Material from patches of thrush should be examined after being stained by Gram's method; budding yeast cells mixed with long filaments, pseudohyphae, are typical and the contrast in size with any bacteria which may be present leaves no doubt as to the fungal nature of the large Gram-positive yeast cells.

Cultivation. The specimen should be plated out on Sabouraud's medium and specimens of sputum or faeces from cases of suspected deep candidiasis should also be inoculated on to a malt tellurite agar and a penicillin-streptomycin blood agar since these selective media help to control the growth of bacterial species which will also be present.

Gram-stained film of material taken from a patch of thrush in a baby's mouth showing *Candida albicans* as long segments of Gram-positive hyphae and oval yeast-like forms. Gram-positive cocci are also present which emphasises the differences in size of bacteria and fungi. ×1000

C. albicans gives large, cream-coloured colonies after 2 to 3 days incubation but cannot readily be differentiated from other species of *Candida* although colonial appearances on differential media, e.g., tetrazolium or eosin-methylene blue can provide a presumptive identification and allow the selection of colonies for further study.

In the diagnostic laboratory confirmation of species identity is usually obtained by subculture into a series of sugars and noting the pattern of acid and gas production, but it is known that the fermentation reactions of *Candida* species are not constant.

Other more specific tests to identify *C. albicans* are available and one of these, the pseudo-germ tube test, deserves wider use; the test is simple, rapid and reliable and requires only the inoculation of a predominantly yeast-forming culture of *C. albicans* into undiluted human or animal serum. The mixture is examined microscopically after 2 hours incubation at 37°C for the presence of pseudo-germ tubes emerging from the yeast cells.

There is an increasing awareness that species other than *C. albicans* can cause not only oral and vaginal thrush but are also incriminated in deep-seated infections; of these species, *C. stellatoidea* and *C. tropicalis* are most frequently involved and the detailed identification of these and other species is usually undertaken in Reference Laboratories.

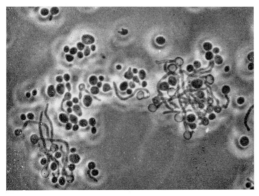

Serum tube test. Cylindrical outgrowths, pseudo-germ tubes, can be seen emerging from the rounded yeast-like cells of *C. albicans*. This unstained preparation was photographed after *C. albicans* had been incubated in serum at 37°C for 1 hr. ×500

CRYPTOCOCCUS NEOFORMANS

Cryptococcus neoformans exists saprophytically in the soil and on vegetable matter and although it is occasionally recovered from the faeces of healthy human beings it is not normally a symbiont of man. It causes cryptococcosis in man, an infection which occurs occasionally in Britain; although lesions occur on the skin and in the lungs, the organism has a particular predilection for brain tissue giving rise to subacute or chronic meningitis.

Infection may on occasion be endogenous but it is more commonly acquired exogenously by inhalation and preceding the onset of meningitis there is a transient and self-limiting lung infection.

C. neoformans is a yeast belonging to the same family as *Candida* and reproduces by budding but does not produce hyphae or pseudohyphae and is characteristically surrounded by a thick capsule.

Laboratory diagnosis depends primarily on microscopic examination of cerebrospinal fluid or other specimens to detect the characteristic spherical, capsulate yeast cells which measure from 5 to 20μ in diameter.

Material should also be inoculated on to Sabouraud's medium and incubated at 37°C for at least 14 days; colonies may be apparent within a few days and young cultures resemble those of *Staphylococcus albus* but colonies later become tan coloured and then brown. Colonies rapidly develop a mucoid character and India ink films of cultures show the characteristic capsulated and budding yeast cells with no evidence of a mycelium.

C. neoformans is not readily differentiated from all other cryptococci on the basis of biochemical activities but it is the only species which is pathogenic for mice; within a week or two of intraperitoneal inoculation of a culture the animal dies and postmortem examination reveals gelatinous masses of *C. neoformans* in the abdomen and the brain.

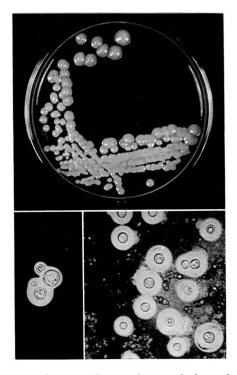

Cryptococcus neoformans. The top photograph shows the appearance of colonies of *C. neoformans* after growth on Sabouraud's medium for 8 days at 37°C; at an earlier stage of growth the colonies resembled those of *Staph. albus* but are now heavily mucoid and would later become tan coloured. $\times \frac{1}{2}$

Of the two India ink films that on the left was made from a centrifuged deposit of cerebrospinal fluid (C.S.F.) taken from a young girl suffering cryptococcosis whilst on the right the preparation is of material taken postmortem from the brain of a mouse 3 weeks after it had been inoculated intraperitoneally with the C.S.F. ×600. The presence of budding, thick-walled spherical cells with prominent capsules is diagnostic.

BIBLIOGRAPHY

CHAPTER 1

BARER, R. (1968). *Lecture Notes on the Use of the Microscope.* 3rd Ed. Oxford & Edinburgh: Blackwell Scientific Publications.

CRUICKSHANK, R. (1968). *Medical Microbiology.* 11th Ed. (revised reprint). Edinburgh & London: Churchill Livingstone.

DUGUID, J. P. (1951). The demonstration of bacterial capsules and slime. *J. Path. Bact.* **63,** 673.

DUGUID, J. P. & GILLIES, R. R. (1957). Fimbriae and adhesive properties in dysentery bacilli. *J. Path. Bact.* **74,** 397.

GILLIES, R. R. & DUGUID, J. P. (1958). The fimbrial antigens of *Shigella flexneri. J. Hyg., Camb.* **56,** 303.

GOULD, G. W. & HURST, A. (1969). *The Bacterial Spore.* London & New York: Academic Press.

KNAYSI, G. (1948). The endospore of bacteria. *Bact. Rev.* **12,** 19.

MILES, A. A. & PIRIE, N. W. (1949). *The nature of the bacterial surface.* First Symposium of the Society for General Microbiology. Oxford & Edinburgh: Blackwell Scientific Publications.

SPOONER, E. T. C. & STOCKER, B. A. D. (1956). *Bacterial anatomy.* Sixth Symposium of the Society for General Microbiology. Cambridge University Press.

TITTSLER, R. P. & SANDHOLZER, L. A. (1936). The use of semi-solid agar for the detection of bacterial motility. *J. Bact.* **31,** 575.

CHAPTER 2

COLLEE, J. G., WATT, B., FOWLER, E. B. & BROWN, R. (1972). An evaluation of the GasPak system in the culture of anaerobic bacteria. *J. appl. Bact.,* **35,** (1), 71.

FRY, B. A. & PEEL, J. L. (1954). *Autotrophic micro-organisms.* Fourth Symposium of the Society for General Microbiology. Cambridge University Press.

HEWITT, L. F. (1950). *Oxidation-reduction Potentials in Bacteriology and Biochemistry.* 6th Ed. Edinburgh & London: Churchill Livingstone.

McINTOSH, J. & FILDES, P. (1916). A new apparatus for the isolation and cultivation of micro-organisms. *Lancet,* **1,** 768.

SALTON, M. R. J. (1964). *The Bacterial Cell Wall.* Amsterdam: Elsevier.

WILLIAMS, R. E. O. & SPICER, C. C. (1957). *Microbial ecology.* Seventh Symposium of the Society for General Microbiology. Cambridge University Press.

CHAPTER 3

ACKROYD, J. F. (1964). *Immunological methods.* A symposium organised by C.I.O.M.S. Oxford & Edinburgh: Blackwell Scientific Publications.

ELEK, S. D. (1948). The recognition of toxicogenic bacterial strains *in vitro. Brit. med. J.* **1,** 493.

OUCHTERLONY, Ö. (1949). An *in vitro* test of the toxin-producing capacity of *Corynebacterium diphtheriae. Lancet,* **1,** 346.

PEETOOM, F. (1963). *The Agar Precipitation Technique and its Application as a Diagnostic and Analytical Method.* Edinburgh: Oliver & Boyd.

WEIR, D. M. (1972). *Immunology for Undergraduates.* 2nd Ed. (reprint). Edinburgh & London: Churchill Livingstone.

CHAPTER 4

COWAN, S. T. & STEEL, K. J. (1961). Diagnostic tables for the common medical bacteria. *J. Hyg., Camb.* **58,** 357.

CHAPTER 5

ANDERSON, E. S. & WILLIAMS, R. E. O. (1956). Bacteriophage typing of enteric pathogens and staphylococci and its use in epidemiology: A review. *J. clin. Path.* **9,** 94.

CADNESS-GRAVES, B., WILLIAMS, R., HARPER, G. J. & MILES, A. A. (1943). Slide-test for coagulase-positive staphylococci. *Lancet,* **1,** 736.

ELEK, S. D. (1959). *Staphylococcus pyogenes and its relation to disease.* Edinburgh & London: Churchill Livingstone.

FISK, A. (1940). The technique of the coagulase test for staphylococci. *Brit. J. exp. Path.* **21,** 311.

WILLIAMS, R. E. O., BLOWERS, R., GARROD, L. P. & SHOOTER, R. A. (1966). *Hospital infection: causes and prevention.* 2nd Ed. London: Lloyd-Luke (Medical Books) Ltd.

CHAPTER 6

FULLER, A. T. (1938). The formamide method for the extraction of polysaccharides from haemolytic streptococci. *Brit. J. exp. Path.* **19,** 130.

LANCEFIELD, R. C. (1933). A serological differentiation of human and other groups of hemolytic streptococci. *J. exp. Med.* **57,** 571.

MAXTED, W. R. (1948). Preparation of streptococcal extracts for Lancefield grouping. *Lancet,* **2,** 255.

THOMAS, C. G. A. & HARE, R. (1954). The classification of anaerobic cocci and their isolation in normal human beings and pathological processes. *J. clin. Path.* **7,** 300.

CHAPTER 7

BOWERS, E. F. & JEFFRIES, L. R. (1955). Optochin in the identification of *Str. pneumoniae. J. clin. Path.* **8,** 58.

DOWNIE, A. W., STENT, L. & WHITE, S. M. (1931). The bile solubility of pneumococcus with special reference to the chemical structure of various bile-salts. *Brit. J. exp. Path.* **12,** 1.

LOGAN, W. R. & SMEALL, J. T. (1932). A direct method of typing pneumococci. *Brit. med. J.* **1,** 76.

CHAPTER 8

GORDON, J. & MCLEOD, J. W. (1928). The practical application of the direct oxidase reaction in bacteriology. *J. Path. Bact.* **31,** 185.

CHAPTER 9

LINELL, F. & NORDÉN, Å. (1954). *Mycobacterium balnei:* A new acid-fast bacillus occurring in swimming pools and capable of producing skin lesions in humans. *Acta tuberc. scand.* Suppl. 33.

MACCALLUM, P., TOLHURST, JEAN C., BUCKLE, G. & SISSONS, H. A. (1948). New mycobacterial infection in man. *J. Path. Bact.* **60,** 93.

MARKS, J. & RICHARDS, M. (1962). Classification of the anonymous mycobacteria as a guide to their significance. *Mon. Bull. Minist. Hlth Lab. Serv.* **21,** 200.

REPORT (1962). Anonymous mycobacteria in England and Wales. 1.—Prevalence and methods of identification. P.H.L.S. *Tubercle,* **43,** 432.

REPORT (1963). BCG and Vole bacillus vaccines in the prevention of tuberculosis in adolescence and early adult life. *Brit. med. J.* **1,** 973.

CHAPTER 10

ANDERSON, J. S., COOPER, K. E., HAPPOLD, F. C. & MCLEOD, J. W. (1933). Incidence and correlation with clinical severity of *gravis, mitis,* and intermediate types of diphtheria bacillus in a series of 500 cases at Leeds. *J. Path. Bact.* **36,** 169.

OUCHTERLONY, Ö. (1949). An *in vitro* test of the toxin-producing capacity of *Corynebacterium diphtheriae. Lancet,* **1,** 346.

CHAPTER 11

KEPPIE, J., HARRIS-SMITH, P. W. & SMITH, H. (1963). The chemical basis of the virulence of *Bacillus anthracis. IX.* Its aggressins and their mode of action. *Brit. J. exp. Path.* **44,** 446.

CHAPTER 12

MACLENNAN, J. D. (1943). Anaerobic infections of war wounds in the Middle East: Bacteriology and Pathology. *Lancet,* **2,** 94.

NAGLER, F. P. O. (1939). Observations on a reaction between the lethal toxin of *Cl. welchii* (type A) and human serum. *Brit. J. exp. Path.* **20,** 473.

PARKER, M. T. (1969). Postoperative Clostridial infection in Britain. *Brit. med. J.,* **3,** 671.

WILLIS, A. T. (1964). *Anaerobic Bacteriology in Clinical Medicine.* 2nd Ed. London: Butterworth.

WILLIS, A. T. & HOBBS, G. (1959). New media for the isolation and identification of clostridia. *J. Path. Bact.* **77,** 511.

BIBLIOGRAPHY

CHAPTER 13

EDWARDS, P. R. & EWING, W. H. (1962). *Identification of Enterobacteriaceae.* 2nd Ed. Minneapolis: Burgess.

EDWARDS, P. R. & FIFE, M. A. (1952). Capsule types of *Klebsiella. J. infect. Dis.* **91,** 92.

GILLIES, R. R. (1956). An evaluation of two composite media for preliminary identification of Shigella and Salmonella. *J. clin. Path.* **9,** 368.

GILLIES, R. R. (1964). Colicine production as an epidemiological marker of *Shigella sonnei. J. Hyg., Camb.* **62,** 1.

HOLMAN, R. A. (1957). The use of sulphonamides to inhibit the swarming of *Proteus. J. Path. Bact.* **73,** 91.

CHAPTER 14

GABY, W. L. & HADLEY, C. (1957). Practical laboratory test for the identification of *Pseudomonas aeruginosa. J. Bact.* **74,** 356.

GILLIES, R. R. & GOVAN, J. R. W. (1966): Typing of *Pseudomonas pyocyanea* by pyocine production. *J. Path. Bact.* **91,** 339.

GOVAN, J. R. W. & GILLIES, R. R. (1969). Further studies in the pyocine typing of *Pseudomonas pyocyanea. J. Med. Microbiol.,* **2,** 17.

KOVACS, N. (1956). Identification of *Pseudomonas pyocyanea* by the oxidase reaction. *Nature (Lond.),* **178,** 703.

MACPHERSON, J. N. & GILLIES, R. R. (1969). A note on bacteriocine typing techniques. *J. Med. Microbiol.,* **2,** 161.

CHAPTER 15

GHAN, K. H. & TJIA, S. K. (1963). A new method for the differentiation of *V. comma* and *V. El Tor. Amer. J. Hyg.* **77,** 184.

MONSUR, K. A. (1963). Bacteriological diagnosis of cholera under field conditions. *Bull. Wld Hlth Org.* **28,** 387.

MUKERJEE, S. & GUHA ROY, U. K. (1962). The vibrios of the recent cholera-like outbreak in Hong Kong. *Brit. med. J.* **1,** 685.

CHAPTER 16

DAVIS, D. J. (1921). The accessory factors in bacterial growth. IV. The 'satellite' or symbiosis phenomenon of Pfeiffer's bacillus (*B. influenzae*). *J. infect. Dis.* **29,** 178.

SMITH, C. H. (1954). Relation between haemagglutination and pathogenicity in strains of *Haemophilus* isolated from the eye. *J. Path. Bact.* **68,** 284.

CHAPTER 17

LACEY, B. W. (1954). A new selective medium for *Haemophilus pertussis,* containing a diamidine, sodium fluoride and penicillin. *J. Hyg., Camb.* **52,** 273.

CHAPTER 18

TALBOT, J. M. & SNEATH, P. H. A. (1960). A taxonomic study of *Pasteurella septica,* especially of strains from human sources. *J. gen. Microbiol.* **22,** 303.

CHAPTER 19

HUDDLESON, I. F. & ABELL, E. (1928). Behaviour of *Brucella melitensis* and *abortus* toward gentian violet. *J. infect. Dis.* **43,** 81.

PICKETT, M. J. & NELSON, E. L. (1955). Speciation within the genus *Brucella.* IV. Fermentation of carbohydrates. *J. Bact.* **69,** 333.

CHAPTERS 20-22

SWAIN, R. H. A. (1955). Electron microscopic studies of the morphology of pathogenic spirochaetes. *J. Path. Bact.* **69,** 117.

—— (1957). The electron-microscopical anatomy of *Leptospira canicola. J. Path. Bact.* **73,** 155

CHAPTER 23

BRONER, M. & BRONER, M. (1971). *Actinomycosis.* 2nd Ed. Bristol: J. Wright & Sons Ltd.

PORTER, I. A. (1953). Actinomycosis in Scotland. *Brit. med. J.* **2,** 1084.

SULLIVAN, H. R. & GOLDSWORTHY, N. E. (1940). Comparative study of anaerobic strains of *Actinomyces* from clinically normal mouths and from actinomycotic lesions. *J. Path. Bact.* **51,** 253.

BIBLIOGRAPHY

CHAPTER 24
WATSON, K. C. (1954). Clot culture in typhoid fever. *J. clin. Path.* **7**, 305.

CHAPTER 26
ELLIOTT, W. A. & SLEIGH, J. D. (1963). Container for transport of urine specimens at low temperature. *Brit. med. J.* **1**, 1142.

O'GRADY, F. & BRUMFITT, W. (1968). *Urinary Tract Infection.* London: Oxford University Press.

TURNER, G. C. (1961). Bacilluria in pregnancy. *Lancet*, **2**, 1062.

CHAPTER 27
COOK, G. T. & JEBB, W. H. H. (1952.) Isolation of ß-haemolytic streptococci from nose and throat swabs by aerobic and anaerobic incubation. *Mon. Bull. Minist. Hlth Lab. Serv.* **11**, 18.

HOLMES, M. C. & LERMIT, A. (1955). Transport and enrichment media in the isolation of haemolytic streptococci from the upper respiratory tract. *Mon. Bull. Minist. Hlth Lab. Serv.*, **14**, 97.

RUBBO, S. D. & BENJAMIN, M. (1951). Some observations on survival of pathogenic bacteria on cotton-wool swabs. Development of a new type of swab. *Brit. med. J.* **1**, 983.

WILLIAMS, R. E. O. (1958). Laboratory diagnosis of streptococcal infections. *Bull. Wld Hlth Org.* **19**, 153.

CHAPTER 28
RAWLINS, G. A. (1953). Liquefaction of sputum for bacteriological examination. *Lancet*, **2**, 538.

CHAPTER 29
ARMSTRONG, E. C. (1954). The relative efficacy of culture media in the isolation of *Shigella sonnei. Mon. Bull. Minist. Hlth Lab. Serv.* **13**, 70.

McNAUGHT, W. & STEVENSON, J. S. (1953). Coliform diarrhoea in adult hospital patients. *Brit. med. J.* **2**, 182.

RUBBO, S. D. & BENJAMIN, M. (1951). Some observations on survival of pathogenic bacteria on cotton-wool swabs. Development of a new type of swab. *Brit. med. J.* **1**, 983.

THOMAS, M. E. M. (1954). Disadvantages of the rectal swab in diagnosis of diarrhoea. *Brit. med. J.* **2**, 394.

CHAPTER 31
LITTLE, L. A. & DEVINE, J. M. (1958). The effect of centrifugation on the cultivation of tubercle bacilli from sputum. *Mon. Bull. Minist. Hlth Lab. Serv.* **17**, 239.

SAXHOLM, R. (1958). *An Experimental Investigation of Methods for the Cultivation of* Mycobacterium tuberculosis *from Sputum.* Oslo: Oslo University Press.

THOMAS, C. H. H. (1956). The laryngeal swab method of diagnosing pulmonary tuberculosis. *Mon. Bull. Minist. Hlth Lab. Serv.* **15**, 141.

CHAPTER 32
HUCKSTEP, R. L. (1962). *Typhoid Fever and other Salmonella infections.* Edinburgh & London: Churchill Livingstone.

WATSON, K. C. (1954). Clot culture in typhoid fever. *J. clin. Path.* **7**, 305.

CHAPTER 33
ALSTON, J. M. & BROOM, J. C. (1958). *Leptospirosis in Man and Animals.* Edinburgh & London: Churchill Livingstone.

REPORT (1956). Diagnosis and typing in Leptospirosis. *Wld Hlth Org. techn. Rep. Ser.* **113**.

CHAPTER 34
MACLEAN, I. H. (1937). A modification of the cough plate method of diagnosis in whooping cough. *J. Path. Bact.* **45**, 472.

SPINK, W. W., McCULLOUGH, N. B., HUTCHINGS, L. M. & MINGLE, C. K. (1952). Diagnostic criteria for human brucellosis. *J. Amer. med. Ass.* **149**, 805.

SPINK, W. W. (1956). *The Nature of Brucellosis.* Minneapolis: Lund Press Inc.

STUART, R. D., TOSHACK, S. R. & PATSULA, T. M. (1954). The problem of transport of specimens for culture of gonococci. *Canad. J. publ. Hlth*, **45**, 73.

WILLCOX, R. R. (1961). Perspectives in venereology—1960. *Bull. Hyg. (Lond.)*, **36**, 825.

CHAPTER 35

GARROD, L. P. & O'GRADY, F. (1971). *Antibiotic and Chemotherapy*. 3rd Ed. Edinburgh & London: Churchill Livingstone.

KAVANAGH, F. (1963). *Analytical microbiology*. New York: Academic Press.

MAXTED, W. R. (1953). The use of bacitracin for identifying Group A haemolytic streptococci. *J. clin. Path.* **6**, 224.

REPORT (1961). Standardization of methods for conducting microbic sensitivity tests. *Wld Hlth Org. techn. Rep. Ser.* **210.**

CHAPTER 36

DUBOS, R. & DUBOS, J. (1952). *The White Plague*. London: Gollancz.

EVANS, D. G. (1969). *Immunization against Infectious Diseases*. *Brit. Med. Bull.*, **25**, (2).

FELIX, A., RAINSFORD, S. G. & STOKES, E. J. (1941). Antibody response and systemic reactions after inoculation of a new type of T.A.B.C. vaccine. *Brit. med. J.* **1**, 435.

HILL, A. B. & KNOWELDEN, J. (1950). Inoculation and Poliomyelitis: A statistical investigation in England and Wales in 1949. *Brit. med. J.* **2**, 1.

HOME, F. (1759). Medical Facts and Experiments. London.

MCCLOSKEY, B. P. (1950). The relation of prophylactic inoculations to the onset of poliomyelitis. *Lancet*, **1**, 659.

REPORT (1957). Field and laboratory studies with typhoid vaccines. *Bull. Wld Hlth Org.* **16**, 897.

REPORT (1963). BCG and Vole bacillus vaccines in the prevention of tuberculosis in adolescence and early adult life. *Brit. med. J.* **1**, 973.

SYMPOSIUM ON IMMUNIZATION IN CHILDHOOD (1959). Edinburgh & London: Churchill Livingstone.

WESTWATER, M. L. (1965). Spacing immunizations. *Brit. med. J.* **1**, 503.

WILSON, G. S. (1967). *Hazards of Immunization*. University of London: Athlone Press.

WOOD, A. (1855). New method of treating neuralgia by the direct application of opiates to the painful points. *Edinb. med. J.* **82**, 265. (Discovery of hypodermic syringe.)

CHAPTERS 37-41

FRENKEL, J. K., DUBEY, J. P. & MILLER, J. N. (1970). *Toxoplasma gondii* in cats: faecal stages identified as coccidian oocysts. *Science, N.Y.*, **167**, 893.

GARNHAM, P. C. C. (1966). *Malaria parasites and other Haemosporidia*. Oxford & Edinburgh: Blackwell Scientific Publications.

HUTCHISON, W. M., DUNACHIE, J. F., SIIM, J. CHR. & HALLEMAN, B. L. (1970). Coccidian-like nature of *Toxoplasma gondii*. *Brit. med. J.*, **1**, 142.

ROBERTSON, D. H. H., LUMSDEN, W. H. R., FRASER, K. F., HOSIE, D. D. & MOORE, D. M. (1969). Simultaneous isolation of *Trichomonas vaginalis* and collection of vaginal exudate. *Brit. J. vener. Dis.*, **45**, 42.

SABIN, A. B. & FELDMAN, H. A. (1948). Dyes as microchemical indicators of a new immunity phenomenon affecting a protozoan parasite (*Toxoplasma*). *Science, N.Y.*, **108**, 660.

WRIGHT, F. J. & BAIRD, J. P. (1968). *Tropical Diseases*. 3rd Ed. Edinburgh & London: Churchill Livingstone.

CHAPTERS 42-44

BENHAM, R. W. (1957). Species of candida most frequently isolated from man: methods and criteria for their identification. *J. chron. Dis.*, **5**, 460.

RIDDELL, R. W. (1951). Laboratory diagnosis of common fungus infections. In *Recent Advances in Clinical Pathology*. 2nd Ed. S. C. Dyke, p. 77. London: Churchill.

TASCHDJIAN, C. L., BURCHALL, J. J. & KOZINN, P. J. (1960). Rapid identification of *Candida albicans* by filamentation of serum and serum substitutes. *Am. J. Dis. Child.*, **99**, 212.

WINNER, H. I. & HURLEY, R. (1966). *Symposium on Candida Infections*. Edinburgh & London: Churchill Livingstone.

INDEX

239